RAVES FOR ADRIAN BERRY'S SPACE ODYSSEY!

"A fascinating account of one of the most bizarre developments of modern science. Thoroughly enjoyable for every point of view."
—Dr. Robert Jastrow
Goddard Space Flight Center, NASA

"Berry's book has already caused a stir in England. Astoundingly he proposes and describes the actual making of some future society of a pair of black holes inside our Galaxy . . . instantaneous space-travel can be achieved by maneuvering through these awesome man-made 'time machines.' . . . few who are *au courant* with the newest paradoxes of astrophysics will fail to be riveted by his lucid, painstaking descriptions . . . Exciting reading with excellent diagrams." —*Publishers Weekly*

"Historians of the 3rd millenium may find Mr. Berry's book a fascinating example of predictive imagination." —Gerard K. O'Neill
Professor of physics at Princeton University and author of *The High Frontier*

"Exhilarating reading."
—*Kirkus Reviews*

"This splendidly imaginative book is going to take the bread out [illegible]
writers."

THE IRON SUN

Crossing the Universe Through Black Holes

Adrian Berry

A Warner Communications Company

WARNER BOOKS EDITION

Library of Congress Catalog Card Number 76–52318

ISBN 0–446–89796–5

This Warner Books Edition is published by arrangement with E. P. Dutton, Inc.

Cover design by Gene Light
Cover art by John Berkey
Warner Books, Inc., 75 Rockefeller Plaza, New York, N.Y. 10019

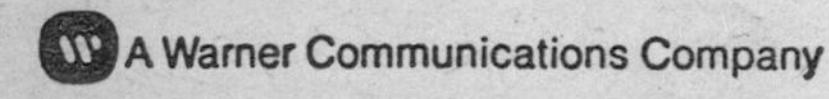

Printed in the United States of America

Not associated with Warner Press, Inc. of Anderson, Indiana

First Printing: November, 1978

10 9 8 7 6 5 4 3 2 1

To My Mother

Contents

Acknowledgments

A great many people havc helped me to write this book. Foremost among these was Academician Anthony Lawton, chief engineer of a prominent European electronics company, a Fellow of the British Interplanetary Society, and like myself, a Fellow of the Royal Astronomical Society. He read the manuscript through twice with the minutest attention, corrected my errors, and with untiring patience, during discussions which sometimes lasted until 3 a.m., helped me to work out some of the concepts in science and engineering which I could at first only dimly grasp. Without his help, the writing of this book would have been impossible. However, the responsibility for any errors which remain uncorrected is entirely my own.

I am particularly grateful also to Dr Anthony Michaelis for reading the manuscript twice and making many useful suggestions, even while he was in the throes of launching a new scientific magazine. I thank my wife and my parents for doing the same. I had most helpful discussions and correspondence with Professors John A. Wheeler, James Bardeen, Kip S. Thorne, David Robinson, P. C. W. Davies, Gerard O'Neill, Derek Lawden, Roger Penrose, John Taylor, Robert L. Forward, Carl Sagan, W. H. McCrea, Eric Laithwaite, F. J. Tayler, Christopher Gregory and P. G. Manning. I must stress, however, that none of these people are to be held responsible for any part of the contents of this book.

Grateful thanks are due also to Patrick Moore, Alan Bond, Ian Ball, Enda Jackson, Clare Dover, John Delin, Robin Birkenhead, Don Milne, George Evans, Kenneth Gatland, Nicholas Craze, Eleanor Berry, Linda Kelly, Lawrence Kelly, John Fadum, John Campbell, Ewan MacNaughton, Clem Wood, Jessie Wood and Billy Nitze.

I specially thank Bob Karbowski for drawing the diagrams.

I had much helpful assistance also from the staff at the Science Museum Library, Kensington, the Royal Institution library, the Royal Astronomical Society library, the National Reference Library for Science and Invention, and the London Library.

On the occasions where I have had to give measurements of distance, density and mass in the text of the book these are given in the older imperial measurements, since I judged that they would be more familiar to most readers. However, in the Glossary and Notes, I give them in both imperial and metric units.

Introduction

This book is an attempt to solve what will be one of the most formidable problems of future ages, the feasibility of travel to the stars. Failure to solve it could bring about the eventual stagnation and ruin of the human species. Yet if it can be solved, whether by the method outlined in this book or by any other, the prospect will instead be of the establishment of a Galactic community, a society in which our descendants will be scattered through millions of worlds in orbits around countless stars. The race will be safe for ever from the threat of extinction, and there need be no limit to the flowering of human culture which this diversity will produce.

Why, it may be asked, would people wish to travel to the stars even if it eventually becomes possible for them to do so? The answer lies deep in human history and prehistory. The desire to migrate to somewhere else has always been a powerful social force. There was a time, about 2 million years ago, when our ancestors were confined to certain parts of East Africa. About 1 million years later, there was a considerable community in China. In the last 50,000 years, people spread southwards to Australia and New Zealand, and somehow across the Bering Strait from Asia to North America. The reopening of the New World, from the sixteenth century onwards, brought waves of migration to the Americas which no European government could halt. The belief that there exist countless undiscovered worlds in the vast reaches of Galactic space makes probable the final act of migratory expansion. As the great Russian rocket pioneer Konstantin Tsiolkovsky once remarked, 'Earth is the cradle of mankind, but one cannot live in the cradle for ever.'

The means that will make this expansion possible will be the development of an instantaneous transport system in space; one in which a spaceship can disappear in one place and reappear in almost the same instant in another place, without having passed any point in between. Stated as crudely as this, the idea must sound insane. But I have written this book to convince others, as several distinguished scientists are already convinced, that the possibility of such journeys follows remorselessly from fundamental law, and that they will one day be a practical proposition.

New mechanical devices and new knowledge have often moved peoples more effectively than ideologies and passions. Many writers, for instance, have discussed the reasons why Europeans migrated into little explored parts of the world in the sixteenth and seventeenth centuries to build colonial empires. Their main motive was obviously the desire for trade and wealth. But this explanation does not really answer the question, since people have always desired to increase their wealth. A more accurate answer lies in the invention of the one mechanical device which made their migrations possible: the ocean-going sailing ship.

This book is about a type of vehicle and a system of transport that is likely, during the next millennium, to become the equivalent in space of the sailing ship. Space is the only future arena for the migrations of mankind. This has been inevitable since 1957, when the first satellite was launched, and perhaps for long before that.

Since 1957, satellites of far greater sophistication have come to serve an immense variety of human needs. Twelve men have walked on the Moon, and it is likely that the dawn of the twenty-first century will see permanent Lunar colonies. More recently, it has been suggested by a distinguished professor of physics, Gerard O'Neill of Princeton, that cities could be built in space that would house tens of thousands of people, built with materials to be taken from the surface of the Moon.[1] Space, with its abundance of raw materials, is as much a part of our environment as a wheatfield or a coal mine; and it is this fact which makes nonsense of the claims that there can exist any final limit to the extent of our economic growth.

Imaginative plans already exist for the further colonization of our local Solar System, as I reported in some detail in a recent work.[2] The planet Venus is likely to have its unbreathable carbon dioxide atmosphere replaced by a more congenial one of oxygen and nitrogen by the introduction of algae micro-organisms. Flying city states, housing tens of thousands of people, with the cylindrical design predicted by Professor O'Neill, will orbit the Earth and Sun in countless sizes and numbers.

The stars are millions of times further from us than the furthest of our planets; a journey to them in any conventional type of spaceship could take centuries. Instantaneous travel will be the ultimate answer to this problem. Yet this will not merely be a question of designing a new form of space vehicle; it will demand in addition a gigantic engineering project in space, which is far more ambitious, both in scale and in cost, than any of the schemes I have so far listed. Because of the much greater expense that it will demand and the scientific knowledge that will be needed to carry it out, it seems impossible that it could ever become feasible during the next century or even in the first half of the twenty-second century. On present trends, it will be possible in the early twenty-third. The phrase 'present trends' is intended as a roughly accurate statistical projection which combines the three factors of economic, technological and scientific growth.

Economic growth is fairly easy to measure and predict by extrapolation from the past provided that one uses a sufficiently long time-scale. The present wealth of the human race, the 'gross world product', was estimated to have a value of about 5 trillion U.S. dollars (5,000,-000,000,000) in 1974, the latest year for which figures were available at the time of writing.[3]

But the gross world product is far from being a fixed quantity. In 'real' terms—that is, after adjusting for inflation—this wealth has tended to grow for a very long time throughout the world at an annual average rate of about 3 per cent.[4] Three per cent each year may not seem very much when one is considering the next decade, but when we contemplate the effects of this growth over a time-scale of centuries, predictions soon become startling. A British politician, R. A. Butler, a man with some knowledge of these effects of compound interest, predicted in 1950 that the standard of living in Britain—the amount of goods which a wage packet would buy—would have doubled in value by 1975. His political opponents laughed at him, and assumed he was simply electioneering for his party, which was then in power. But it happened! He knew that Britain's national income per head was increasing by 2.8 per cent a year, as it had done for more than a century, and he foresaw the effects which this would have. Anything which increases at 2.8 per cent a year at compound interest must double every twenty-five years. His sceptical opponents, who were thinking in terms of simple interest, could not understand this; to Butler, who understood compound interest, it was obvious.

Let us consider therefore what humanity's $5 trillion will become in the next few centuries. I will assume an annual growth rate of 3 per cent. This is perhaps a somewhat conservative figure, but it is essential to be cautious and allow for recessions. Growing at 3 per cent, this wealth will double about every twenty-three years.* So far, in view of the British example which I have just

*This is *not* a short-term forecast. All I am saying is that the average annual economic growth rate of humanity, in real terms and over an extremely long time-scale, is likely to be 3 per cent. There will always be local variations. At the time of writing, for instance, West Germany's economy is growing at an annual rate of 6 per cent and Britain's is hardly growing at all.

quoted, this does not sound very dramatic. But as time goes forward, the effects of compound interest become extraordinary. Bearing in mind that the following dates are intended to be approximate, the growth will be something like this. If humanity's wealth was $5 trillion in 1974, it will have reached $10 trillion before the year 2000. Table 1 shows how it might continue to grow (with the dates and figures slightly adjusted for simplicity).

Table 1

Year	*Gross human wealth (trillions of dollars)*
2025	20
2050	40
2080	80
2100	160
2125	320
2150	650
2170	1300
2190	2600
2215	5000

These figures, I emphasize again, are 'real' in that they have been adjusted for inflation. They assume also that the resources of space, the material resources to be found *outside* the atmosphere of this planet, will be as readily exploitable as the surface resources of the planet are today. It is thus a firm prediction that someone in the year 2215, in the first quarter of the twenty-third century, will stand a reasonable chance of being *a thousand times richer* than his counterpart in the same relative social and economic bracket would be today.

But this growth of wealth cannot come without a corresponding increase in knowledge and its applications. It is insufficient to define standard of living simply as the amount of goods which a wage packet—or a corporation, or a government—can buy. The *quality* of those goods must also be taken into account. This demands a continuous growth in technology and science, a growth which today continues at a pace which, viewed by the standards of any past era, would be considered unbelievably rapid. Only with this growth of knowledge can mankind acquire the physical energy to increase his wealth and extend his power over nature. Examples from the past are just as dramatic as the invention of the sailing ship; the bow and arrow enabled man to kill at a distance and dominate all other animals. By the superiority of iron swords over bronze ones he subdued his enemies and built nation states. The U.S.A. built its strength in the nineteenth century with the telegraph and the railways. Modern trade is made possible by the aircraft, communication by the satellite, and complex administration by the computer, while relative peace appears to be assured for the present by the existence of nuclear bombs. For this growth to continue indefinitely, it is necessary only that it shall continue to be matched by the growth of science.*

One piece of knowledge will grow increasingly important as time goes forward. There will come a time when every businessman and administrator will need to be as familiar with the effects of Albert Einstein's Special Theory of Relativity as he is familiar with global time-zones today. Many readers may wince at this, and their alarm is understandable. A friend of mine, a quite unscientific person, was studying at Harvard a few years ago, and found that he was required to take at least one science subject.

'I chose astronomy,' he told me.

> I wasn't sure what it was, but it sounded rather exciting, and I thought it would have something to do with telling one's fortune by the stars. The first afternoon, some as-

*A nuclear war can delay, but cannot halt this process, as I explained at least to my own satisfaction in *The Next Ten Thousand Years*. Other scientific studies have since predicted that a substantial part of the human race would survive even a full-scale nuclear war. Progress would be set back by anything between 50 and 500 years. But these are small time-scales compared with the estimated future lifetime of the Sun of some 6,000 million years.

tronomy lecture notes were placed in front of us, and they were covered with horrible equations, all about relativity. My God, I nearly blew my lunch!

This extreme reaction was unnecessary; for many laymen can, with a small effort, understand the Special Theory far more easily perhaps than they can grasp much more difficult concepts of finance and economics. There is no need for equations. The thing can be explained in only a few sentences. I will endeavour now to explain it, so that much that follows in this book will be easily comprehensible.*

Light in the vacuum of space moves at an unchanging speed of 670 million m.p.h., *irrespective of the speed of its source.* This means that if an astronaut is racing through space directly towards a star at half the speed of light he might reasonably imagine that the light from that star will be reaching him at the combined speed of one and a half times the speed of light.

And so he gets out his most accurate clock to see whether this is true.

To his astonishment, he finds that it is not. The light from the star ahead is still reaching him at 670 million m.p.h. After checking and rechecking, there is only one conclusion that he can logically draw from that fact: *his clock must be running slowly.* In other words, time on board the spaceship has slowed down.

This is one of the main predictions of the Special Theory; that time always slows down in a moving vehicle. The faster the vehicle is moving, the more does time inside it slow down. And 'time' does not merely mean what is recorded by the astronaut's clock; it means the actual ageing of the spacecraft and the ageing of the astronaut's body. It is no exaggeration to say that if he travels through space fast enough, say at 99.9 per cent of the speed of light, he can return to Earth and find that he is physically younger than his own children who

*For a fuller explanation of the two main theories of relativity, please see Appendix III.

have stayed behind! And this is all because, for a still unknown reason, the speed of light is constant, irrespective of the speed of its source. If the ship is going even faster, even more extreme consequences result. If it were to be somehow accelerated to a velocity of 99.9999 per cent of the speed of light, the astronaut could return after a journey of, say, five of his years to find that the Earth had aged *millions of years* in his absence. It is idle to speculate which time is the 'right' time and which the 'wrong' time; both are right. Time does not proceed everywhere at a constant speed. Its rate of passage depends on the speed through the Universe of the clock that is measuring it.*

The Special Theory makes other important predictions, as will be seen later. In particular, it shows that no spaceship, nor indeed any material object, can ever exceed or even reach the speed of light itself in this Universe. As it approached that speed, for reasons similar to those which slow down time, the length of the spacecraft, from nose to stern, would become steadily shorter, until at the speed of light itself, if that were ever possible to attain, it would become zero. This ship, in other words, would cease to be a material object; it would have no more material existence than the image of a photograph. The nearest star, the nearest sun to our own, is so far away from us that even if we were travelling at the speed of light we could not reach it at anything less than 4⅓ years. The speed of light therefore represents an impassable barrier to expansion. A journey through the vast reaches of interstellar space is restricted absolutely by this speed limit.

*One might legitimately ask: speed in relation to what? Surely the Earth is moving also through space, and its speed should be taken into account. Einstein's answer is that the spaceship is moving in relation to the whole Universe, and the Earth is not. When the Earth moves, the whole universe moves with it, and when the spaceship moves it moves alone. And so the spaceship's time changes in relation to the time of any 'fixed' object like the Earth.

This book describes a method by which this supposedly impassable barrier can be penetrated, and how *instantaneous* journeys in space can be achieved—as opposed to a faster than light journey which can never be. It was necessary in this Introduction to outline some aspects of the Special Theory, in the simplest possible form, so that what follows would be intelligible. It was necessary also to show the scale on which human wealth is likely to increase during the coming centuries, for an instantaneous journey through space will require colossal amounts of energy. It is true that man will need only to supply a small fraction of this energy himself, for nature will do the rest. But even that small fraction of this energy would cost a considerable fraction of the world's present wealth, a fraction which is more money than he can afford in this century or the next.

The remainder of the necessary energy, that remainder that nature herself will be able to provide when the time comes and which man will harness to his aid, will be no less than the gravitational energy of ten Suns.

Part One

THROUGH THE BLACK HOLE

What plain proceeding is more plain than this?
SHAKESPEARE, *Henry the Sixth*

1
Where Matter Vanishes

Our bodies are part of the Universe. The very atoms that compose them were forged in the nuclear furnaces of giant stars. New suns were formed from the debris of titanic stellar explosions that took place thousands of millions of years ago. These new stars flung out great streams of incandescent gases and other material which cooled and became planets such as ours. Elements congenial to life reacted in the crust and gaseous atmosphere of our world, and primitive life came into existence.

Now, some 3,000 million years after the coming of the first algae and protozoa, the ultimate ancestors of life, we have evolved into creatures who are beginning to develop space travel. We seek once again to show that we are part of the Universe that created us. This book is about a special kind of space travel. It does not concern the purely local manned expeditions, or the unmanned missions of mechanical instruments that we have sent between the planets of this tiny Solar System. I mean travel to the stars, to other planets in orbit around those remote suns, whose distance from us is so great that even light, travelling at the stupendous velocity of 670 million m.p.h., takes 4⅓ years to reach us from the nearest one and 80,000 years from the furthest in our great Galaxy of 180,000 million stars.

These numbers might seem terrifyingly remote from ordinary life. They astound our imagination even while they are not beyond our comprehension. In a colour photograph of the Galactic star-fields taken through any of the world's largest optical telescopes, we see a myriad tiny points of light, awesome in their numbers, each one of them a sun, and many with a high probability of having habitable planets in orbit around them. Occupation of these planets, and mastery of the immense spaces between them, is the ultimate aim of space travel.

Our achievements of colonizing the Solar System will be but temporary expedients, occupying human energies perhaps for slightly more than the next century and a half. It is the conquest of the Galaxy, the traversing of truly cosmic distances to exploit the great riches of other planetary systems, that will be the enterprise of our descendants from about the beginning of the twenty-third century onwards. Extremely rough estimates, based on the flimsy astronomical data now available to us, suggest that the number of habitable planets in the Galaxy—that is to say, Earth-type worlds with atmospheres of oxygen, nitrogen and water—may approach 500 million.[1] Even assuming that this figure is over-optimistic by as much as 99 per cent, this would still leave us a potential human empire of 5 million worlds.

Yet in an emotional sense, the Galaxy seems as vast and as unexplorable today as the distant parts of the world must have seemed to the people of medieval Europe in the days before long-distance sailing ships; when the very idea of a journey that crossed the Equator and rounded the southern tip of Africa was ruled out by the gloomy predictions of Ptolemy. Ptolemy calculated correctly that the Sun's heat must strike the regions bordering the Equator at right-angles. He wrongly deduced that the temperature must therefore be so great at the Equator that wooden ships entering that region would catch fire.[2] Thirteen hundred years elapsed between Ptolemy's purely mathematical discovery and the daring voyages instigated by Prince Henry the Navigator of Portugal, who had the boldness to send ships across the Equator and prove that Ptolemy had drawn a wrong conclusion from a correct calculation.

It would be a tragedy if we were compelled again to endure to our detriment so long a period between a brilliant discovery being made in mathematical physics and a second discovery that the conclusions being drawn from the first were erroneous. Yet most of those today who seriously consider the possibilities of interstellar communication and travel are in this stage of groping between the first discovery and the second. Einstein's Special Theory of Relativity is a work of transcendent genius that dwarfs the crude calculations of Ptolemy. The Special Theory, as I pointed out earlier, forbids any journey through the Universe that exceeds or even matches the speed of light. At speeds equalling that of light, the theory goes on to predict, the length of a spacecraft would shrink to zero from the viewpoint of an external observer, and the engine-power required for its propulsion would become infinite. Now since no engine *of infinite power* could ever be built, and since a spacecraft of zero length and the people in it could enjoy no more real existence than the images on a photograph, journeys as fast or faster than light appear upon a first consideration to be absolutely impossible.

This conclusion might seem to be valid, not merely today when spacecraft cannot reach even a thousandth of the speed of light, but in all future ages, no matter what powerful and sophisticated engines are constructed. A round-trip journey of nearly nine years to the nearest star or 160,000 to the furthest in our Galaxy would be respectively inconvenient and intolerable for those who waited through long ages for the return of the mission. The people of Spain had to wait three years for the return of Ferdinand Magellan, the first circumnavigator of the Earth; and when his crew returned without him, found it difficult to remember who he had been or when and to what destination he had set out. Still more extreme situations could be predicted in the future, in which astronauts return from journeys of age-long duration, to find an Earth which has changed far beyond their recognition, and in which they can find no records of their departure or even of their existence. It is idle to recall Einstein's prediction, which has been confirmed by experiment, that the speeding astronauts would age far more slowly than their 'stationary' friends back on Earth.[3] It would still be a grim experience for a traveller who had completed an interstellar voyage at velocities near the speed of light in, say, ten of his years to return to a home planet which was 1,000 years older than it was when he left it, and in which all memory of his existence had utterly passed away.

It is possible to see a dim historical parallel between Einstein and Ptolemy. Both men produced theories which, if regarded as being applicable in every detail to all conceivable circumstances, will lead to erroneous conclusions. The equations of Einstein's Special Theory have been seen to represent a comprehensive view of the Universe; they appear to form the absolute basis for modern physical science.[4] Yet they do not; there are regions in space where the known laws of physics break down absolutely. So complete is this breakdown that the very physical constants cease to be either physical or constant. We reach regions where the concept of distance is eliminated, and where the words 'space' and 'time' consequently change their meanings. I refer to the little-understood regions beyond the inner horizons of those phenomena which we call rotating black holes.

No challenge or criticism is intended in any part of this book against Einstein's theories. Their more bizarre-seeming predictions, such as the slowing down of time in a fast-moving vehicle and the actual curvature of space in the presence of a large mass, have been confirmed by experiments whose validity cannot be doubted. But one must quarrel with one particular *interpretation* of Relativity which many people, even a large number of scientists, have made after a superficial or a secondhand study of the equations. Their error is to maintain that the equations of the Special Theory apply in all conceivable circumstances. At learned seminars, for example, on the possibilities of communicating with other intelligent civilizations in the Universe, one is apt to hear statements like:

'Einstein has shown that speeds faster than light are impossible. Therefore a journey through the Galaxy would take centuries.' Einstein indeed said this, both verbally and in writing, but *it is not what his equations say.*

There is a difference between his verbal and his mathematical descriptions of space. It is a subtle difference, and until the early 1960s, it would have seemed quite meaningless. Yet if correctly understood, it will induce an incalculable change in the future history of the human race. A physical equation, like a legal statute, must be interpreted with absolute literalness if it is to be properly understood. A scientist, or for that matter a legislator, may think that he knows what he has written when he establishes a new scientific principle or a law which permits or forbids various actions. But he may be dismayed to find afterwards that the wording of his equation, or law, being rigorous, can permit all sorts of unintended exceptions or 'loopholes', a fact to which many a tax-lawyer owes his living.

So it is with Einstein's key equation which forbids faster than light journeys. The equation, when interpreted literally, does *not* forbid faster than light journeys, as so many people imagine that it does. It only forbids them *in this Universe;* that is to say, in the immediate surroundings in which the spaceship finds itself. The theory neither permits nor forbids astronauts from taking short-cuts through other universes or other physical states which have different physical laws. It has nothing to say on the matter, for the good reason that back in 1905 Einstein had no information with which to speculate about such matters. Yet this is the very situation we now face when we consider the challenge of the black hole.

Black holes represent the ultimate crisis of physics. It has in the past been difficult to predict what happens inside them because, in the words of John A. Wheeler of Princeton, one of the world's leading cosmologists, all known tools of physical calculation become useless, and 'smoke comes out of the computer'. Yet they are important to cosmology, he explains, because they teach us that Einstein's equations are 'purely local in character; they [the equations] tell us nothing about the topology of the space with which one is dealing'.[5]

What are black holes, and what have they to do with topology? Why, in other words, should they be of any use as a cosmic transport system? The astronomical theorist Kip Thorne of the California Institute of Technology describes them like this:

> Of all the conceptions of the human mind from unicorns to gargoyles to the hydrogen bomb, perhaps the most fantastic is a black hole: a hole in space with a definite edge over which anything can fall and nothing can escape; a hole with a gravitational field so strong that even light is caught and held in its grip; a hole that curves space and warps time.
>
> Like the unicorn and the gargoyle, the black hole seems much more at home in science fiction or in ancient myth than in the real Universe. Nevertheless, the laws of modern physics virtually demand that black holes exist. In our Galaxy alone there may be millions of them.[6]

It has taken two centuries for the human mind really to grasp this conception, and even today there are scientists who cannot bring themselves to accept the existence of black holes. When first proposed in 1798 by the Frenchman Peter Simon Laplace, they were fairly simple objects; namely, stars so gigantic and consequently with such huge gravitational fields that nothing, not even light, could escape from them. They would therefore be invisible. This invisibility, or rather the inability of light to escape from them, is the first elementary fact one must learn about a block hole.* Every celestial body, no matter how small or how massive, has what we call an 'escape velocity' which is derived from its radius and mass. This

*I should say that at least one black hole has been identified, probably beyond doubt. It lies about 6,000 light-years from us, in the constellation of Cygnus the Swan, and it orbits a giant star called HDE 226868. It has been identified as a black hole because of its intense emission of X-rays.

is the speed which a vehicle must attain to escape from the body's immediate gravitational pull and reach orbit. As everyone familiar with space rockets knows, the Earth's escape velocity is about 25,000 m.p.h. If anyone succeeded in launching an orbital vehicle from the surface of a more massive body, such as the planet Jupiter, he would find that its comparable escape velocity would be just over 135,000 m.p.h. If his vehicle failed to attain that speed as it soared upwards, it would fall back to the surface.

Now a black hole is far more massive than either the Earth or Jupiter. It is the burned-out hulk of a star which was at least three times more massive than the Sun. Its hydrogen fuel will be exhausted, and it will have radiated light only for about 600 million years. (This is an extremely short life for a star; when the Sun was that age the Earth was scarcely formed. As a general astronomical rule, the more massive a star, the shorter its life.) Less massive stars of this age are liable to blaze out in supernova explosions. But a star of three Solar masses or more is too heavy to explode. Instead, it does the opposite. It begins to collapse under its own colossal weight. The collapse, or *implosion*, begins slowly, but proceeds ever more swiftly. Its radius becomes ever smaller. The escape velocity mounts in proportion, and at length exceeds the speed of light. At this point, since light cannot in any circumstances surpass its normal velocity of 670 million m.p.h., neither light nor anything else can escape from the former star. It has become a black hole.

I have so far written nothing which should be difficult to believe. This is the simplest possible description of a black hole. But the situation is far more extraordinary than this. The collapse of matter will be halted at various stages, depending on the size of its original mass. It is necessary to outline these stages briefly, so that what follows will be intelligible.

A star of the Sun's mass will end its long life (in the Sun's case about 6,000 million years from now) as a dim-shining, condensed solid body about the size of the

Earth, called a *white dwarf*. A piece of material the size of a sugar cube from a white dwarf star would weigh about five tons.* Some time after the identification of the white dwarf Sirius B, nine light-years from Earth, early this century, many astronomers simply refused to believe that this was possible. As Sir Arthur Eddington wrote in 1926:

> The message of the companion of Sirius when decoded ran: 'I am composed of material 3,000 times denser than anything you have come across. A ton of my material would be a little nugget you could put in a matchbox.' What reply can one make to such a message? The only reply which most of us made in 1914 was: 'Shut up. Don't talk nonsense.'[7]

It was clear by 1926 that Sirius B was not 'talking nonsense', and was indeed transmitting a correct description of itself. But a star more massive than Sirius B, a star with an original mass more than 1.4 times the Sun's, will not stop contracting at the white dwarf stage.† Its gravitational collapse will continue until its very atoms have destroyed one another, leaving only the extremely dense neutral particles, or neutrons, giving the name *neutron star* to the fantastically compressed object that remains. A neutron star is about ten miles across, and so dense that a sugar cube-sized piece of material from it weighs about 100 million tons! Yet this is scarcely surprising when we remember that a mass of one and a half Suns has been compressed into a body no larger than Greater London.

A star of more than three Solar masses cannot even survive as a neutron star. It is precisely at this stage in the death-throes of a star of this mass that its escape velocity reaches the speed of light, and it collapses with catastrophic suddenness into a black hole. Then, very slowly, tak-

*Some white dwarfs are believed to consist of extremely densely packed crystalline carbon, or pure diamond.

†This figure of 1•4 Solar masses is known as 'Chandra's limit', after the Indian astronomer who first calculated it in 1931, Subrahmanayan Chandrasekhar.

ing several million years in the process, the black hole begins to crush itself out of existence. Not even the neutrons can now escape being destroyed. The star's volume becomes ever smaller. As the diameter shrinks, density rises towards infinity. The ten-mile diameter becomes five miles, then one mile, and then a few hundred yards. Still the shrinkage continues. The diameter becomes a few hundredths of an inch. Soon it is no more than the width of an atom. Shortly after this, it becomes smaller than the smallest sub-atomic particle, and after that so small that no instrument devised by man or any other intelligent creature could have any chance of measuring it. Roger Penrose, of London University, has speculated on what would happen if an extremely rugged instrument were lowered into a black hole while this process of slow contraction was going on. (He is of course only making a 'thought experiment', since the experiment would be pointless because of the instrument's inability to send back any radio message or television picture describing the conditions which it found.) Once the instrument has passed beyond the black hole's *event horizon,** communication with the outside world ceases. It has reached the point at which no signal could be sent back to the outside world because to do so it would have to exceed the speed of light. From that moment on, the instrument is doomed:

> As it falls towards the centre of the black hole [writes Penrose], the mountaing tidal forces will rise rapidly, ripping to pieces in turn the material of the instrument, the molecules of which this instrument is composed, the atoms which constitute those molecules, the atomic nuclei, and finally the fundamental particles which a moment ago had been the building blocks of those nuclei. And the entire process would not last more than a few thousandths of a second!†[8]

*The event horizon, as its name clearly suggests, is the point beyond which no communication with the outside world is possible. Please see Glossary for further information.

†The black hole takes several million years to crush itself out of

This is an almost unbelievable situation. Not only does a massive star slowly contract until it vanishes altogether out of this Universe, taking its mass with it by achieving infinite density; it is also creating in space a *singularity,* a region where the density of matter is so high that the curvature of space becomes infinite, and where, consequently, any object, no matter how ruggedly constructed, is stripped to the point where its fundamental particles no longer exist and it vanishes completely.*

This 'curvature' of space is a bewildering idea when one first encounters it. In 1916, Einstein produced what was probably his greatest work, the General Theory of Relativity; it was far more subtle and complex than his earlier Special Theory. In summary, it shows that planets orbit the Sun because the Sun's mass causes space to curve or warp in the area surrounding it. The Sun's mass creates a ring-like corridor—there is no better word for it—whose walls are made of curved space, and the Earth travels eternally through this corridor. All the celestial bodies in the Universe are confined to their orbits in similar curved gravitational corridors created by the masses of larger bodies, and that, in crude essence, is all one needs to know for the purposes of this book about the General Theory of Relativity.†

In the same year that Einstein produced the General

---

existence; yet the instrument it utterly destroyed in a few thousandths of a second. There is no contradiction here. The black hole is several times the mass of the Sun, and the instrument is presumably a small, man-made object. The enormous difference in mass between the two objects dictates the respective time-scales in which they are destroyed.

*It can be deduced from this that two black holes which were formed at the same time from equal quantities of mass will be indistinguishable. No possible test could ascertain what material they were composed of. Hence the saying, 'a black hole has no hair'. You can't tell whether a bald man's parents were blond or brunette.

†For a fuller explanation of the General Theory of Relativity, please see Appendix III.

Theory, a German officer named Karl Schwarzschild, a scientist in civilian life, contracted an infectious disease on the Russian front. He was invalided home to Berlin. He had only a few weeks to live, but he spent this period producing two of the most brilliant scientific papers of the age. He studied Einstein's prediction of the curvature of space, and carried this prediction to its most extreme conclusion: namely, that when matter exceeds three times the mass of the Sun the curvature of space around it eventually becomes infinite. He thus calculated the properties of black holes as I have described them above. He found the precise diameter to which any body of known size and mass would need to be contracted before its collapse became inexorable. He discovered that this point is reached when the diameter of a star of Solar density shrinks to approximately 0.0005 per cent of its original size. The Sun, for instance, would become a black hole if its present diameter of 865,000 miles was, by some miracle of compression, shrunk to just under four miles. Beyond that point, nothing could prevent eventual collapse to a size that was infinitesimally small. And absolute contraction would lead not only to the absolute annihilation of all the Sun's matter, but to the equal annihilation of all other matter that fell into the consequent black hole.

Einstein's prediction of curvature, detectable in the curvature of light-rays, was confirmed by Eddington's famous photographs of the changed positions of stars during a Solar eclipse in 1919, which proved that starlight curved in the presence of a large mass. Enormous publicity resulted. Schwarzschild's so-called 'solution' to Einstein's equations seemed to have been equally confirmed by Eddington's experiment, and there the matter rested for forty-four years. People simply lost interest in gravitational theory, not because the subject was uninteresting but because, as Freeman Dyson of Princeton explains, 'nobody could think of any new observations or experiments'.[9] Gravitational effects on the tiny scale of the atom were too weak to be measured, and nothing new was observed in the heavens that was sufficiently bizarre to require the General Theory to explain it. A few bold men, like J. Robert Oppenheimer and John A. Wheeler continued to write with enthusiasm of the untapped scientific riches of gravitational collapse, but observational astronomers tended to ignore them; in the words of the British theorist Brandon Carter, 'they brushed aside collapsed objects as figments of our imagination'.*[10]

Then, in 1963, came the startling discovery of quasars, or quasi-stellar objects. Nobody, to this day, has been able to decide exactly what quasars are, but their very existence was enough to destroy the complacency of those who thought they knew everything about the sky. Briefly, quasars are objects apparently at the very edge of our Universe, only a few thousand times bigger than the Sun, but which shine with the brilliance of 100 galaxies; that is to say that they radiate as brightly, when we allow for their extreme distance from us, as 10 trillion stars! Nobody could explain such gigantic emissions of energy in terms of the orderly, fairly placed Universe which had been in fashion before 1963. The astronomers were forced to admit that forces existed in nature of which they understood nothing, and the observatories and universities buzzed with excited talk about Schwarzschild collapses, supernovae explosions in chain reactions, matter and antimatter in collision, and all sorts of violent and extraordinary phenomena. The ferment of this period was all the greater since it followed a long age of dullness. I interviewed many astronomical theorists at this period, while working on a cover story about the Universe for *Time* magazine. The experience was almost like watching the last act of the *Götterdämmerung*. Everything was 'catastrophic' and 'cataclysmic'. One was left with a vague feeling of relief that one was still alive, but that this happy situation could not last, for there seemed to be

*It is hard to think of two scientists whose views it was more reckless to ignore than Oppenheimer and Wheeler. The two men had formidable achievements to their credit; Oppenheimer's team detonated the first atomic bomb, and Wheeler was a co-inventor of the hydrogen bomb.

scarcely a body in the heavens that was not exploding, or was about to explode, with ungovernable fury.

At the height of the excitement in 1963, there appeared in one of the scientific journals a modest paper of one and a half pages which none the less seems destined to affect our knowledge even more profoundly than the discovery of quasars. Its author was a quiet New Zealander named Roy P. Kerr, then at the University of Texas at Austin, and his subject was the nature of black holes.[11] Like the announcements of many great scientific breakthroughs, the publication of Kerr's paper passed wholly unnoticed by the general public, and little observed, at least for a year or so, by the quasar-happy astronomers. But it may be no exaggeration to suggest that in future ages, Kerr's discoveries will seem as important in human history as the invention of the wheel. His work, and the belated endorsement of it by other scientists of distinction, represent, at any rate, the principal reason why I have dared to write this book. For not the least of his predictions is that Schwarzschild's belief in the inevitable annihilation of *all* matter that falls into a black hole is wholly wrong.

2
The Spinning Gateway

The inquiring mind often suffers from too much specialization. Nearly two generations of scientists explored the Schwarzschild solution in the utmost detail without ever asking themselves the central question: Does it represent a reasonable hypothesis in principle? Is it likely, in other words, that black holes remain motionless and do not rotate? For this absolute stillness is what Karl Schwarzschild assumed. Whether he really believed that any object in the Universe could exist without rotating, we may never know; but certainly he did not incorporate rotation into his equations.

Every object in the cosmos, however, rotates on its axis at one speed or another. Planets, stars and galaxies all rotate. It is even possible that our entire Universe rotates, although whether this is true, or at what speed, or in relation to what, there seems to be at present no way of ascertaining. Bodies in our small Solar System rotate at more or less leisurely speeds. The fastest rotating body is Jupiter, on whose surface a whole day and night would last just under ten hours on an Earthly time-scale.

Yet a super-dense body, such as a neutron star, will rotate at a very high velocity because of its very great density, that is to say its very large mass contracted into a very small volume. A neutron star, it will be remembered from the last chapter, is an object slightly less massive than a black hole, and which, although incredibly compressed, does not undergo complete gravitational collapse. Yet it spins on its axis at a tremendous rate. When the first neutron star was discovered in the crab Nebula in 1967, it was actually thought for some weeks to be the source of a signal from an intelligent civilization in space. The reason why radio astronomers did not rule out this possibility is obvious; the neutron star was emitting radio bleeps just over thirty times each second with the accuracy of an atomic clock. Such precision seemed quite unnatural, and the astronomers at once suspected they were receiving signals from 'little green men'. The theorists, after thinking for a few months, assured them that there was a natural explanation: the neutron star was emitting radio 'pulses' as it rotated, and the greater the density of an object in space, the more rapidly it must rotate. The frequency of the radio pulses from this neutron star was simply indicating its very high speed of rotation.* To put this another way, if we were standing on the surface of a neutron star (assuming that such an object could have a 'surface' as we understand the term), we would find that each day and night would last one thirtieth of a second!

We have seen how the central region of a black hole is almost infinitely dense.† Consider how much more rapidly it must therefore rotate than a neutron star. It has been computed that a black hole of ten Solar masses will rotate at speeds in the region of 1,000 times per second, so that the time equivalent of a twenty-four-hour period on Earth would last no more than a thousandth of a second

*The appropriate term 'pulsar' was invented to describe a neutron star by Dr Anthony Michaelis, former science correspondent of the London *Daily Telegraph,* and now editor of the journal *Interdisciplinary Science Reviews.* The term is now widely used and is synonymous with 'neutron star'.

†It is at least 10^{92} pounds per cubic inch (about 10^{94} grams per cubic centimetre). The former figure implies no less than 1 followed by 92 zeros! The differential factor of density between that of intergalactic space (10^{-31} grams per cubic centimetre) and the singularity of the black hole is thus 10^{123} (1 followed by 123 zeros) the biggest number I have ever heard of emerging from natural science.

on the 'surface' of a spinning black hole. This is more than twenty times the speed of rotation of an old-fashioned aircraft propeller.

Rotation was the great change which Roy Kerr introduced into black hole theory in 1963. He invited the world to regard black holes as rapidly spinning objects.* Until then, those few scientists who retained an interest in the matter had laboriously considered them as if they were stationary. Kerr's suggestion produced an understanding of the black hole that was far more rich and complex than anything envisaged by those who had studied the solution of Schwarzschild.[1] Moreover, it has since been proved that Kerr's analysis was correct, as I shall explain in a moment; all black holes rotate, and the non-rotating black holes which Schwarzschild envisaged cannot exist in nature.[2] Today, as we shall see, as a consequence of Kerr's ingenuity, rotating black holes are coming to be regarded as the actual navigable gateways through which spaceships in future centuries will be able to make instantaneous journeys to other parts of the Galaxy.

To discover the real meaning of the Kerr solution, we must consider rotation as the agent of that powerful source of energy known as centrifugal force, where matter is hurled outwards. Anyone who has made himself ill by visiting the 'wall of death' at a fairground will know something of centrifugal force. People paying for that masochistic pleasure stand around the wall of a large cylinder. A lever is pressed and it rotates, slowly at first and then faster. They are pressed to the wall, unable to move. To make the experience more dramatic, the floor is at the same time lowered. Instead of dropping with it, the victims are transfixed by the centrifugal force as if they were so many flies.

Consider now a much more extreme case of the effects of centrifugal force. Let us suppose that an imaginary

*This was not Kerr's first intention as he started his mathematical analysis. As he said afterwards: 'I wasn't looking for a spinning solution; it just came out of the equations.'

scientist, Dr Malevolent, wishes to be revenged on a large group of his colleagues who have constantly denigrated his work. He decides to murder them all, and being of a bloodthirsty nature, he wants to kill them with machine-gun fire. His task is difficult, because he wants the mass murder to look like an accident.

At length Dr Malevolent comes up with a plan for machine-gunning people without using a machine-gun. He invites his colleagues to a lecture (no doubt with free drinks as a bait) in a windowless classroom with concrete walls. On the stage is the murderous professor standing beside a lead sphere about two feet in diameter, and a steel rod through its centre fixing it to the floor.

'I wish to demonstrate some curious effects of the rotating Kerr black hole,' he explains:

> This lead ball is a poor substitute for black hole material but it's the best I can find. At least it is heavier and denser than most terrestrial materials. You will observe the steel rod that fixes the ball to the metal floor of this stage. This rod must represent the axis of the black hole, around which it spins. I have had the floor of this stage specially strengthened with several tons of cast iron, and believe me the floor will need it. The rod is connected to an electric motor whose power must substitute for the law of conservation of angular momentum. You are surely familiar with this law? It is as old as Newton. It predicts simply that once an object in the vacuum of space is spinning, its rate of spin cannot be reduced except by the application of some counter-force. Very well, what happens? As I press this switch, so, you see the rod starting to rotate and the ball rotating with it. The electric motor is quite powerful in fact, and with a minimum of friction, within a few minutes the ball should achieve a speed of rotation of a thousand times per second. You should feel some interesting effects. *There is no escape. I have locked the doors!*

So saying, the evil professor disappears through a back exit, which he locks behind him. The ball spins ever faster, its rod-axis at right-angles to the audience. At first nothing else seems to happen, then a perceptible bulge appears around the ball's equator. The bulge becomes more pronounced as the ball spins faster. As the rotation rate reaches several hundred times per second, the ball suddenly fragments. Lumps of lead tear across the room at supersonic velocities. The room becomes a charnel house. The audience has been raked with a hail of lead, many of the 'bullets' ricocheting off the concrete walls, and the effect is precisely as if the room had been sprayed by a machine-gun.[3]

What Dr. Malevolent had realized and the complacent audience had failed to understand was the true gravitational effects of a Kerr black hole. Here is what might appear the strangest of paradoxes. The greater the density, the greater the speed of rotation; but the density is in turn destroyed by this rotation. The contraction of mass without limit creates infinite densities, as Schwarzschild correctly foresaw. But he did not foresee the immensely rapid rotation which these densities cannot fail to produce. All the particles of matter which now comprise the black hole were themselves rotating. By a fundamental law, known as the law of conservation of angular momentum, this rotation by the particles cannot stop, even though they have crashed into a central body. Instead, the sum of their individual frequencies of rotation is transferred to the new central body. Angular momentum cannot be destroyed, and the speed of rotation of the final black hole equals the individual speeds of rotation of all the original spinning particles which now comprise it, all added together. And because a black hole is very densely packed with particles that once were spinning, it may itself rotate at speeds of around 1,000 times per second.*[4]

Consider an ice skater spinning in a pirouette with arms outstretched. Her arms rotate more slowly than her body, because of their distance from her centre of gravity. Suddenly she drops her arms to her sides, and her speed of rotation is at once increased. This has nothing to do with any lessened effects of air resistance; she now spins

*This figure assumes a black hole with ten times the mass of the Sun. A more massive black hole will rotate even faster.

faster because the former velocity of rotation of her arms is now added to that of her body. She will rotate faster even than this if she is carrying heavy weights in her hands. Why? Because again, to conserve angular momentum the rotation of the masses of these heavy weights must also be added to her own rotation.

It is fairly obvious that centrifugal force must bring about a great *decrease* in density in those parts of the black hole where that force is strongest. Centrifugal force, in fact, has the opposite effects of those of gravitational collapse. We saw how Dr Malevolent's victims saw their doom in the growing bulge round the equator of the lead ball. It was the radical decrease in the ball's equatorial density, with every part round the equator striving to break loose from the centre, that made the ball burst into fragments and kill them. The rotating black hole will have a similar bulge at its equator that indicates a stupendous decrease in density compared with its average density overall. Its equatorial regions will be in fact about 10^{30} times (that is, 1 followed by 30 zeros) *less* densely packed with matter, than, on average, is the rest of the black hole.[5]

In short, the faster the rotation, the lower the density around the equator. A black hole should therefore be a strange-looking object—or would be if it was ever possible to photograph one. The equatorial bulge will be so great that the black hole cannot be a sphere at all. It will instead assume the almost flat shape of a lens, or a disc with a bulge at the centre. Better still, imagine about four gramophone records piled on top of one another. Suppose also that in the centre, where there would normally be small holes that secure the records to the turntable of the gramophone, there is a roughly spherical bulge, as if someone had broken the centre of the records and put an apple there instead. Here is the general shape of a rotating black hole.[6]

Even the Earth, with its incomparably slower rotation of once every twenty-four hours, has a very slight likeness to this shape. The Earth is not completely round, as many people suppose. The poles, through which its axis passes, are slightly flattened; and all around the Earth's equator, both in the land masses and along the ocean beds, there is an almost imperceptible outward bulge. These very slight deviations from pure roundness are due to the planet's slow rotation. If the rotation was any faster, the polar flattening and the equatorial bulge would be correspondingly more pronounced. The faster an object spins in space, the flatter its shape becomes. We may see this general law at work in the shape of many of the galaxies, those vast groupings of thousands of millions of stars which rotate together as if they were single entities. In proportion to their immense sizes, they rotate much faster than the Earth. And so often their shapes are flattened until they assume the general form of a lens. But their rotation is in turn much slower than that of a black hole, and so they seldom reach the stage of being flattened into a thin disc.

It may be objected that it is all very well to make assertions of this kind about galaxies, which can be observed and measured through telescopes with the utmost precision, but that it requires a degree of impudence to make equally detailed statements about invisible holes! The answer is that such statements are demanded by physical theory, the same tool of science that predicted the nature of white dwarfs and neutron stars long before the true strangeness of those objects was confirmed by observation. Soundly argued theory has consistently given us accurate visions of the cosmos, and it is to theory that we must turn to discover what actually happens within the inner regions of a Kerr black hole.

The easiest way to consider the possibilities of actual journeys into black holes is to draw the appropriate maps of space and time, or 'Kruskal diagrams'. Dr M. D. Kruskal, of Princeton, developed and refined in 1960 an old device for analysing time and space, by drawing maps of them.[7] A map of time and space might seem a meaningless concept, because both time and space are abstractions. But this need not prevent us from drawing a map of them which means something, provided that a certain period of time is made to correspond to a certain fixed distance in space; such a map will have a real meaning if it describes something real, such as the speed of light. The simplest possible Kruskal diagram is shown in Figure 1.

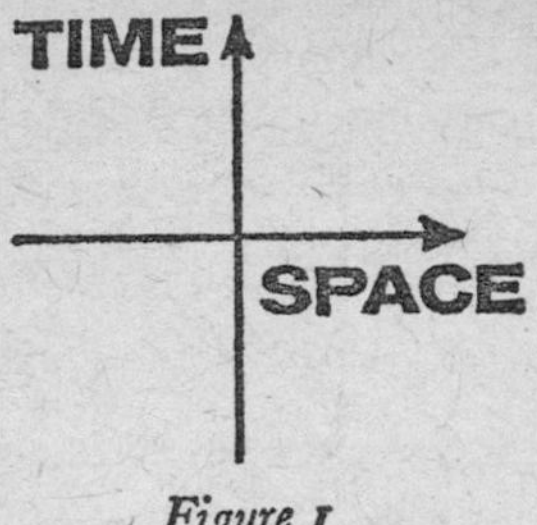

Figure 1

The speed of light, it will be remembered, is 670 million m.p.h. One hour, therefore, on the vertical 'time line' in Figure 1 corresponds to 670 million miles on the horizontal 'space line'. Two hours would correspond to twice that distance: 1,340 million miles; four hours would mean four times the distance, and so on. This makes it possible to elaborate the diagram in a very convenient way, which I will do in a moment. We see that the two lines cross each other at a right-angle, as they do in any graph. If a diagonal line is drawn across the diagram at an angle of 45°, exactly half a right-angle (so as to set out, in the form of a simple graph, the relation of a given time to a given space), this will correspond conveniently to the speed of light.

Now we know that any journey slower than light is 'possible', in the sense that it is not forbidden by any fundamental law. We know also that any journey either at the speed of light or beyond it is impossible. In Figure 2, these three sorts of journeys are illustrated. The journey that is slower than light, and possible, is shown by a diagonal line which crosses the diagram with an angle of more than 45°. The journey at the speed of light is precisely 45°. And the impossible journey that is faster than light has a line crossing the diagram at less than 45°.

So far, all this is very simple. There are three journeys, which I have marked A, B and C. Trip A is possible, and trips B and C are impossible. Two technical expressions are necessary here for later comprehension. Trip A is much nearer to being parallel to the 'time line' than it is

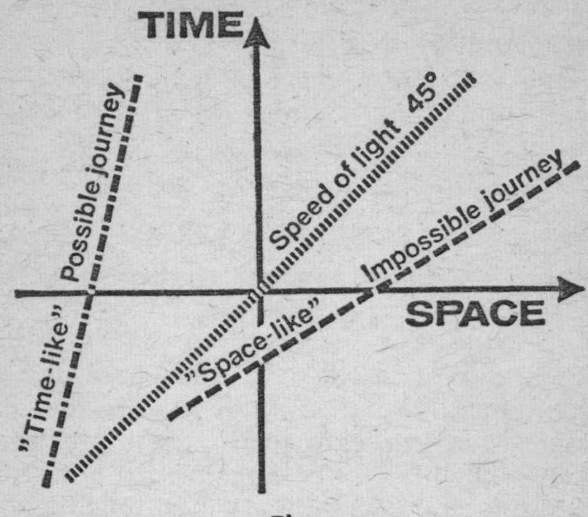

Figure 2

to the 'space line'. It is therefore called a *time-like* journey. Trip C, by contrast, is nearer to being parallel to the 'space line'. We therefore call it a *space-like* trip. And so a time-like journey is always possible, because it is slower than light, while a space-like journey is always impossible, because to carry it out one would need to go faster than light, which is impossible.

The next step is to draw a Kruskal diagram which describes a black hole. Let us start with a hypothetical Schwarzschild, non-rotating, black hole and see what happens if a spaceship is aimed straight at it. As I explained earlier, the black hole is surrounded by an 'event horizon' which traps all objects that approach it within a certain distance. This is the point near a black hole beyond which nothing can escape, because to do so it would need to go faster than light. Also in the Kruskal diagram of any black hole must be shown the position of the fearsome 'singularity', the technical name for that region where gravitational forces reach infinite strength, and where any object, no matter how ruggedly constructed, must be crushed into nothing. Figure 3 shows what happens when the ship approaches the non-rotating black hole.

Entering this kind of black hole could be disastrous. Trip A fails, of course, to pass the event horizon; it is slower than light and achieves nothing. But Trip B passes the event horizon, and crashes into the singularity and the spaceship is totally destroyed. Trip C also passes the event horizon, but would need to be faster than light and is therefore impossible. A choice of journeys that are respectively pointless, suicidal and impossible clearly does not seem a promising prospect for instantaneous interstellar travel. For that is what the hypothetical astronaut inside this spaceship is trying to achieve! Following up a prediction made by Einstein in 1935, he believes that he can vanish into the black hole, and reemerge, an immeasurable fraction of a second later, in a different part of the Universe. The only reason why his plan cannot work is that he has entered the wrong kind of black hole.

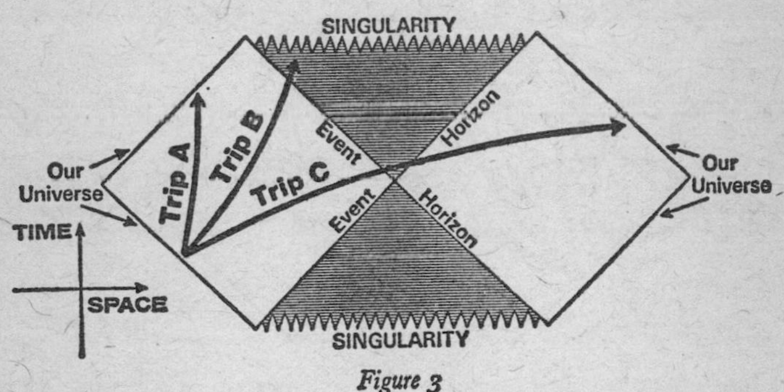

Figure 3

He has entered a kind of black hole which cannot exist. But if the black hole is rotating, if it is a Kerr black hole, the corresponding Kruskal diagram changes completely. It has several surprising properties. In the first place, there are two event horizons, an outer and an inner: the original single event horizon is strangely split into two by the rotation. Second, there is an absolute

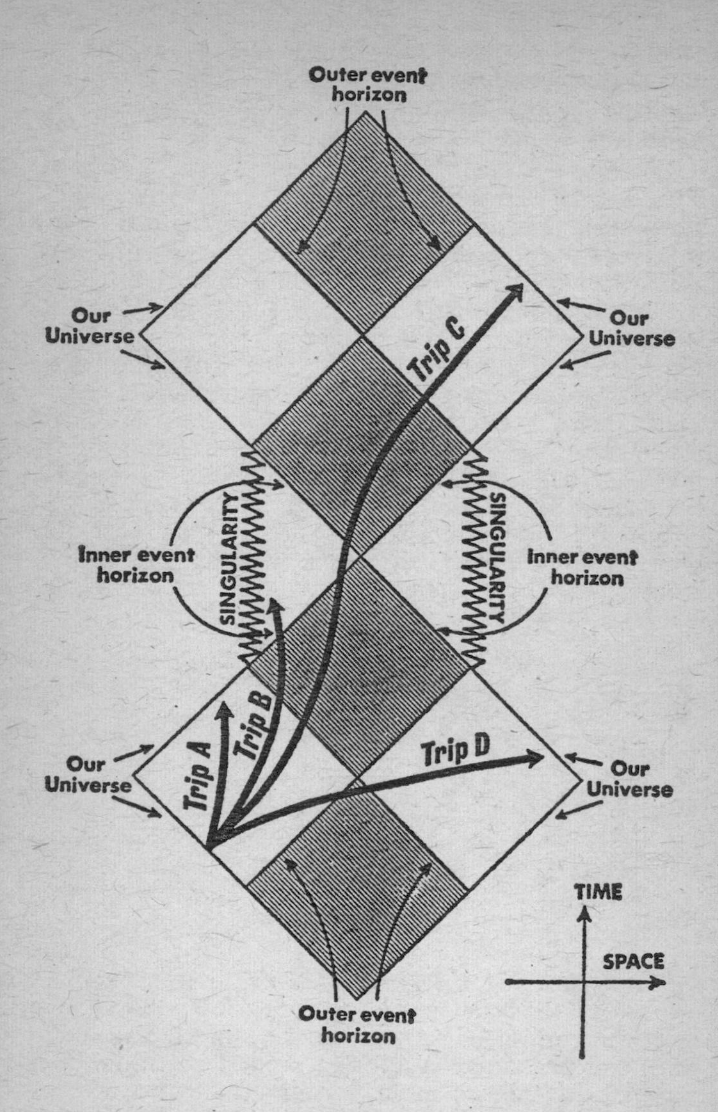
Outer event
horizon
Our
Universe
Our
Universe
Trip C
SINGULARITY
Inner event
horizon
SINGULARITY
Inner event
horizon
Trip B
Trip A
Our
Universe
Trip D
Our
Universe
TIME
SPACE
Outer event
horizon

Figure 4

change in the character of the singularity. Without going into too detailed mathematics, I will say only that as a result of this change the singularity must be aligned upwards with the time axis rather than horizontally with the space axis. This means that the diagram itself must be presented vertically instead of horizontally (see Figure 4—an even more complicated diagram than the previous one, but there is no way of avoiding this).[8]

Now we have four conceivable journeys, Trips A, B, C and D. Trip A fails to pass the event horizon and achieves nothing. Trip B passes the event horizon but hits the singularity. Trips A and B, therefore, are respectively pointless and suicidal. Trip D goes faster than light, and so is 'space-like' and hence impossible. But the astronaut taking Trip C would have the most truly astonishing experience. He passes the outer and the inner event horizons, yet he avoids the singularity. What happens to him then? The prediction made by the diagram is quite plain. He disappears into the black hole and comes out somewhere else. To put it even more bluntly, *he disappears in this region of the Universe and reappears instantaneously in another region of it.* It will be shown in the next chapter precisely how he does this, and why, if his spaceship is aimed correctly, there is no other possible outcome to his journey.

3
Into the Whirlpool

Mr Alvin Toffler's fascinating book *Future Shock* discusses the bewilderment of people whose lives are suddenly affected by social and technological change. This bewilderment can be just as great, in a different way, if the change in question is intellectual rather than material. And so it will come as a considerable shock to many people to realize that physical laws insist that an object in space can disappear in one place and reappear an instant later somewhere else without having passed any point in between. Contrary to what Eastern mystics may say, this is a phenomenon for which there is no previous experience on Earth. Common sense tells us that such behaviour is forbidden; and common sense, so far as the surface of the Earth is concerned, is perfectly right.

But the surface of the Earth is a very sheltered part of the Universe. Nowhere upon it do physical conditions become extreme. Nowhere upon it do gravitational forces become sufficiently strong for bizarre consequences to be observed. Never have ordinary human beings encountered an object like the black hole which, in the words of Kip Thorne, 'curves space and warps time'. Even when the science fiction writers use such literary devices as a 'space warp' to move their spaceships across the Galaxy in no time at all, they are wrongly accused of having resorted to magic in order to make their stories work.

Science fiction writers do not normally study fundamental physics, and their concept of a 'space warp' is arrived at through instinctive feeling rather than knowledge. This is a pity, since the scientific basis for an instantaneous journey, or at least a passage through which such a journey could be made, was worked out in 1935 by no less an authority than Einstein himself.

With his colleague Nathan Rosen he published in that year a paper which is regarded as the forerunner of all modern theories that predict the interconnection of our Universe by timeless passages. Einstein and Rosen used the term 'bridge' to describe this interconnection, and the two men gave a lucid description of an *Einstein–Rosen bridge* that will serve as a model for everything we have been discussing.[1] They likened two separated parts of space to two flat 'sheets' with 'bridges' connecting them. There is a well-known modern analogy which explains the idea of their 'bridge': take a thin tightly stretched sheet of rubber to represent space in one part of the Universe; place on top of it a small wooden ball to represent a planet or a small star; the ball will slightly depress the rubber, making a shallow pit for itself to lie in; the depression in the rubber, caused by the mass of the ball, represents in turn the curvature or warping of space, which is predicted by Einstein's General Theory.

Now take a much heavier ball, say a lead cannonball, to represent a black hole; it will sink deeply into the rubber until its weight tears a hole in it; it will fall through this hole, and the rubber, now freed from its weight, will spring back to its original flat position; suppose that the ball then comes to rest on another flat surface beneath the rubber strip; it is clear what has now happened: the ball, or black hole, has vanished from one part of the Universe and appeared in another; instead of travelling through space in the conventional way, moving along the rubber, it has taken an instantaneous shortcut to another region of space.

This paper was written twenty-eight years before Roy Kerr introduced the concept of fast-rotating black holes. Einstein and Rosen had only the theory of Schwarzschild to work with. And because, as we have seen, any object trying to penetrate a Schwarzschild, non-rotating black hole would be crushed by its singularity, they suspected that the 'bridge' might have no physical meaning, since entrances to it were barred by natural forces.

Roy Kerr has now removed that natural barrier. Another analogy shows what happens when a black hole rotates. Water, under the influence of extreme agitation, behaves in similar ways to space which is warped by the presence of a spinning black hole. Let us see how this works: imagine a man in a small boat where the sea is 300 feet deep. An observer of limited experience in the ways of the sea, a person misled by his own common sense, would say that it was impossible for the man in the boat (if he did not wear a diver's suit) to travel downwards through the water and touch the sea-bed *without getting wet*.

And yet anyone with a knowledge of the vortex effect of opposing tides in a narrow channel can see easily how the man in the boat could make this journey. His boat could be seized in the rotating spiral of a whirlpool. The deep whirlpool in Edgar Allen Poe's terrifying short story *A Descent into the Maelstrom* is an almost perfect terrestrial analogy to a Kerr black hole, with water playing the part of the gravitational fields of space.[2] Consider this passage from Poe, who took the trouble to acquaint himself with the physics of a whirlpool before he wrote his story:

> In a few minutes more, there came over the scene another radical alteration. The general surface of the sea grew somewhat more smooth, while prodigious streaks of foam became apparent where none had been seen before. These streaks, at length, spreading out into a great distance, and entering into combination, took unto themselves the gyratory motion of the subsided vortices, and seemed to form the germ of another more vast.

> Suddenly—very suddenly—this assumed a distinct and definite existence, in a circle of more than a mile in diameter. The edge of the whirl was represented by a broad belt of gleaming spray; but no particle of this slipped into the mouth of the terrific funnel, whose interior, as far as the eye could fathom it, was a smooth shining and jet-black wall of water, inclined to the horizon at an angle of some forty-five degrees, speeding dizzily round and round with a swaying and sweltering motion, and sending forth to the winds an appalling voice, half shriek, half roar, such as not even the mighty cataract of Niagara ever lifts up in its agony to Heaven.[3]

The luckless hero of the story can actually observe a circular or oval area of the sea-bed as his boat spirals down towards it. This area is much smaller than the mile-wide diameter of the top of the funnel, but it is kept temporarily clear of water by the furious rotation of the whirl. But here the analogy ends, for there is no such barrier at the bottom of a Kerr black hole. Instead, the funnel of whirling space goes on—until it opens up again into another region of normal space, whose distance from the point of departure, in ordinary circumstances and in the absence of a black hole, could be as much as several light-years.

The vertical Kruskal diagram of the Kerr black hole, which I showed in the last chapter (Figure 4), is a very useful chart which establishes a principle. But it is very far from being a flight-plan which could guide an astronaut. What should he actually do when he approaches the black hole? Should he just let himself go, abandoning the controls of his spacecraft to the gravitational fields, trusting that he will be safely hurled down the whirling spiral? No, for if he did this he would be sucked into the central singularity and be crushed to pieces. Instead, he must remember the disc shape of the black hole, the shape of about four gramophone records piled tightly on top of each other. This disc (assuming that the black hole has ten times the mass of the Sun) will have a circumference of just under 116 miles. It will be rotating just as a gramophone record rotates on its turntable, but at a velocity of 1,000 complete revolutions each second.

Each part of the disc will therefore be moving at a speed of 116,000 miles per second, or just over 400 million m.p.h.[4] The astronaut must match this speed precisely as he approaches the black hole. This speed, which is slightly more than 60 per cent of the speed of light, will seem tremendously high by the standards of today's space travel. It is in fact nearly 17,000 times faster than the highest speed which the Apollo astronauts attained on their way to the Moon!

Yet this speed is not prohibited by any fundamental law. Exactly *how* it will be attained, whether by such exotic methods as the continuous explosion of matter and anti-matter (please see Chapter 8 for an explanation of this principle), I do not pretend to know. But I have little doubt that the engineers of the far future will solve this problem, just as today's engineers have solved problems that our forefathers would have found incomprehensible.

When the astronaut has matched his speed with that of the spinning disc, each in a sense will now be stationary, relative to each other. A fresh analogy may be helpful here. Imagine a youth standing beside a rotating merry-go-round. He sees an attractive girl riding on one of the wooden horses. He is not content to stand still and admire her each time she comes round. He is determined to watch her continuously as she rides. And so he starts running furiously round and round the merry-go-round until his speed matches precisely the speed of the riding girl. He can then, if he wishes, step on to the platform of the merry-go-round without risk of injury.

So it is with the astronaut and the rotating disc. Each is stationary relative to the other. The astronaut 'looks' at the disc-edge beside him and 'sees' a long rectangular aperture with a height of about 640 yards.* *This aper-*

*Again, the 640-yard height of this aperture assumes a black hole of about ten times the mass of the Sun. It is a black hole of this mass with which we shall later be dealing. The words 'looks' and

ture is the actual gateway to another region in space, and is the one route which passes through both event horizons and avoids the crushing densities of the singularity. Diving directly into the aperture, the astronaut and his ship vanish from the sight of any outside observer. Yet they survive; they vanish only in the sense that a spectator at an airport sees an aircraft and its passengers vanish into the sky. For the astronaut and his ship will have vanished out of this immediate region of space. They will have accomplished the apparent miracle of vanishing in one place and reappearing, a moment later, in another place which may be separated from the point of disappearance by a vast distance. How does this seeming miracle occur? What actually happens in the region between the two inner event horizons in the middle of the Kruskal diagram? The answer is, nothing! The astronaut has reached a place where physical laws, as we know them, end altogether. The region between the inner event horizon of the black hole and the point where the spaceship re-emerges into normal space is the arena for what is called the 'ultimate crisis of physics'.* This region is in many ways highly mysterious, but there is one thing that can be said of it with surety; distances within it are abridged absolutely. The word distance not only loses its present meaning, it ceases to have any meaning. Such units of distance as we know—the light-year, the mile and the inch—are reduced to infinitesimal

---

'sees' here are purely figurative. It is unlikely that the astronaut would 'see' anything at all of the black hole disc outside his spacecraft. He would need his instruments to detect all outside conditions. See also Note 4 to Ch. 3.

*Some scientists term this region 'Superspace', implying that it is a universe much larger than our own, obeying different physical laws. It is believed that Superspace is the universe from which *our* universe emerged when it came into existence some 15,000 million years ago. But these are very deep questions, and for the purposes of this book it is sufficient to understand the principles of space warping around a rotating black hole and the general properties of an Einstein-Rosen bridge.

proportions. This does not just mean that they have become very much smaller; it means instead that they have become so small that they cannot be measured at all, no matter how sophisticated our measuring instruments may be.

It should be said in fairness that not all scientists accept these conclusions. Astronomical theories can sometimes sound so strange that there is a time delay before they are fully accepted, as we have seen in the case of Eddington's comments on the dense white dwarf star. I shall be discussing one such example in the area of instantaneous travel—namely Brandon Carter's suggested revision or even abandonment of Einstein's General Theory—in the next chapter.

But let us return to the disc-edge of the black hole which the astronaut must enter. Turning into it, as he matches the speed of his ship with the speed of the rotating disc, he finds that the density of matter there is surprisingly low. According to a certain mathematical formula, it is not more than about 10^{30} (that is, 1 followed by 30 zeros) *less* than the density of matter in the rest of the hole. This works out as being about 100,000 million times less dense than terrestrial air at sea-level!*

The vertical diameter of the disc-edge, or width of the strip, is about one hundredth of the diameter of the disc of the black hole. This low-density strip corresponds to the bulging equator of Dr Malevolent's lead ball. But the astronaut will be in no danger of being struck by fragments flying out under centrifugal force, as were Dr Malevolent's victims, since the black hole will have an incomparably stronger gravitational field than the lead ball, and ejected matter will therefore escape so slowly that it will present no danger to the spaceship.

Suppose that the black hole in question is ten times more massive than the Sun. Now it can be computed, by

*These density comparisons may sound confusing, and so I have given figures for various different densities, in different parts of a black hole and elsewhere in nature, in the Glossary under 'density'. The formula for calculating the density of matter in the region of the rotating disc is explained in Note 5 of Chapter 2.

means of Schwarzschild's best-known equation, that a black hole of ten Solar masses will have a diameter of just under 37 miles.[5] The navigable aperture will therefore have a diameter of one hundredth of this, which is 640 yards, a gateway which a spacecraft of several thousand tons will surely be able to enter with a reasonable margin of error.

What happens next, after the ship has passed through this strip? We know now what actually occurs, but it is not quite clear why it occurs: by its very act of passing through the strip, the ship has passed through the inner event horizon of the black hole, and it has begun to cross the Einstein–Rosen bridge; an immeasurable fraction of a second later, it emerges in another and distant part of space.[6] I have shown that it does this by entering a region where forward distances are abridged absolutely. Now if we consider this prediction carefully, we see that there is another way of expressing it, which does not change its meaning in the slightest degree. Instead of saying that distances are reduced to the infinitesimal, let us say that time runs backwards.

One layman at least appears to have stumbled upon this transposition independently of the mathematical physicists. The British philosopher Guy Robinson of Southampton University wrote in 1964 a learned and slightly mocking article which poked fun at those science fiction writers who casually describe faster than light voyages, but who never bother to try to explain how they are achieved. Robinson showed, quite accurately, that a person could not travel faster than light because if he did so, apart from other difficulties, he would have to travel backwards in time.[7] In the absence of any special circumstances, at least one of these two feats would plainly be impossible. But Robinson went further. Apparently delighted by the success of this argument, he then expressed the opinion that an instantaneous journey through space would be impossible for the same reason. The only way for an astronaut to travel instantaneously, he suggested humorously, would be for the astronaut to carry a time-machine in his spaceship, so that he could move backwards in time as he moved forwards in space, thus eliminating distance.

It is now certain that Robinson's joke contained a shattering truth. For the Kerr black hole, or for that matter the Einstein–Rosen bridge, *is* a time-machine! A spaceship entering it, provided that it is aimed correctly, is not only flung out in another part of space, but in making this journey it is propelled *backwards* in time. (We shall see in the next chapter that the constraints that normally prevent us from travelling backwards in time do not apply in a black hole.) Several cosmologists find that they can reach only this conclusion when they study the Kerr solution. As Frank J. Tipler of Maryland University puts it, 'a rotating cylinder (i.e. a Kerr black hole) would act as a time machine'.[8] Even Brandon Carter, of Cambridge University, England, who reaches a different conclusion eventually, admits that 'it is possible to connect any event to any other by a future-directed time-like line'.[9] It will be remembered from the Kruskal diagrams that 'time-like' means a legitimate journey that goes slower than light. And so here we have a slower than light journey, that does not violate Einstein's Special Theory, but which nevertheless moves backwards in time, and which for all practical purposes may therefore be regarded as a journey between two separate points which is in effect much faster than light. Yet strictly speaking, because its backwards movement in time corresponds precisely with its forward movement in space, the journey is not faster than light; it is instantaneous.

Now it may seem, with all this talk about time-machines, that I am saying something stranger than anything before. But it is not so; for a time-machine and a distance-abolishing machine are merely two phrases *to describe exactly the same thing*. Distance simply means time travelled, or to be more precise, it is the average speed of a journey multiplied by the time taken to achieve it. For instance, the distance between London and New York is normally covered by a subsonic aircraft at an average speed of 580 m.p.h. in a period of about six hours. Multiply these two figures, and we obtain a true distance between these cities of 3,500 miles. But if this distance of 3,500 miles was miraculously reduced to zero, the aircraft would be able to make the journey instantaneously, in no time at all, because speed multiplied by zero equals zero. Yet to make this instantaneous journey, the aircraft would have to travel backwards in time *while* it was moving forwards in space.

And so Guy Robinson's seemingly fanciful speculation about the use of a time-machine has a very real meaning in the context of an instantaneous journey. We must deal now with some of the awesome problems of logic and philosophy which such journeys would involve.

4
The Forbidden Circle

Professor Brandon Carter is horrified by the conclusions from Einstein's General Theory which I have described in the last chapter. He is so appalled by what he regards as the inevitable consequences of an instantaneous journey that he calls them 'pathological' and 'vicious'. He concludes in a lengthy paper on the problem that the prediction of instantaneous travel, far from being a true description of what could really happen, only serves to reveal a 'very serious breakdown' in the General Theory, and that, consequently, the whole theory may have to be 'abandoned, or at least drastically reformulated'.[1]

Some readers will be relieved at these reflections; thank heavens, they will say, there is at least one sane person in this madhouse. But let us see precisely what Professor Carter accepts and what he finds intolerable. He and several other cosmologists have agreed after detailed studies of the Kerr solution that the possibility of instantaneous communication and travel appears to be a consequence of the General Theory. But their objection is that an instantaneous journey would violate, on two levels, one of the fundamental laws of nature, namely the principle of causality, which states that a consequence cannot occur before the event which caused it. Before explaining why Carter and some of his colleagues believe that causality would be violated, and before seeing whether or not their belief is justified, I must give some account of what causality really is and of some of the strange and impossible situations which a violation of it would produce.

The principle of causality assumes that time can only move in a forward direction. The passage of time is slowed down in a vehicle moving extremely fast, but still it only moves forwards. In the present state of the Universe, there exists only the forward-moving 'arrow of time'. There is nothing in science to suggest that we will ever be able to relive post events. As a poet once lamented:

> Who can undo what time hath done?
> Who can win back the wind?
> Beckon lost music from a broken lute?
> Renew the redness of a last year's rose?
> Or dig the sunken sunset from the deep?[2]

Why do I emphasize so obvious a point? For the very good reason that Einstein's 1905 equations predict that any journey that was faster than light would carry its travellers backwards in time. I have already mentioned other reasons why it would be impossible for a space vehicle to move at the speed of light, let alone beyond it. But let us suppose for a moment that the best-known objections to a faster than light voyage were miraculously disposed of. Imagine that by some magic a spaceship was able to reach and surpass the limiting 670 million m.p.h. and that its length in the direction of motion did *not* shrink to zero, and that the power required for its propulsion did *not* rise to infinity. The consequences for the travellers inside the ship are plain in such a case. They would age progressively more slowly as the ship accelerated. At the speed of light itself, their time would stop altogether, and they would be able to traverse the Universe in what was, for them, one everlasting frozen second. They would have the status of a light beam which, if it were conscious of its existence, would be a cosmic 'Peter Pan', and would cross distances of millions of light-years and never grow older. But at a speed beyond that of light, which, I stress again, is always likely to be impossible, the spaceship and the people inside it would move backwards in time. This is clearly predicted by the Special Theory, by that very same equation in fact which predicts such peculiar phenomena as 'negative length' for a faster than light spaceship.[3] And so, travelling backwards in time, an astronaut who had moved faster than light would be able to return to Earth *before* he set out. There, he could meet a slightly younger version of himself. The two limericks that describe this situation are fairly well known, but they are still most apposite:

> There was a young lady named Bright,
> Who travelled much faster than light,
> She started one day
> In the relative way
> And returned on the previous night.
>
> The lady was Bright but not bright,
> And she joined in next day in the flight;
> So then two made the date,
> And then four and then eight,
> And her spouse got the hell of a fright.[4]

But what would happen if Mrs Bright took a dislike to such high-speed travel after her first voyage, and if, during a conversation with her untravelled 'self', she successfully advised 'herself' that the trip was not worth taking, and that she should phone her travel agent and cancel the ticket. The resulting situation would be ridiculous. The ticket is cancelled and the trip does not take place. 'Self Number 2', who, until that point had been a real human being, not only ceases to exist but *ceases ever to have existed*. The written records that recorded her return to Earth and her transactions since that time are mysteriously expunged. One moment you see her name in a hotel register; the next moment it has gone. The hotel clerk who checked her in and the porter who carried her baggage can remember no such person. They are not lying; the reason why they remember no such person is because there was no such person. There 'was' such a person admittedly, but the new situation, since the cancellation of the ticket, is that she does not exist. The past, the very real past, has been altered with an efficiency which no totalitarian secret service could rival.

This is an illustration of the mess in which we find ourselves when we try to violate the principle of causality. This fairy-tale situation is one of the basic examples of such a violation. Anyone who postulates a faster than light or instantaneous journey or even an instantaneous message is obliged to ask himself the question: Can Jones receive a message from Smith *before* Smith has sent it? If the answer, after a careful examination, was shown to be 'yes', then we would have to affirm that Jones and Smith could not possibly exist and that the whole situation belongs to the realms of fantasy. Why? For the same reason that Mrs Bright cannot duplicate herself. There is a time *after* Jones has received the message during which Smith is at liberty to change his mind and decide not to send it. But he does not in fact have this 'liberty' because Jones has already received the message. Smith therefore *must* send the message, and his free will, his right to do whatever he pleases, has been taken away from him. The 'circle of time' has been closed, and causality has been violated, because Smith has lost his right of free will. Any theory of the Universe, however plausible and sublime it may be in all other respects, is certain to be nonsense if it closes the circle of time.

Consider another impossible case. You are accosted in the street one day by a man who looks exactly like you but is a few years older. He tells you that he is your future self and that he has built a time-machine and visited his past in order to meet you. You think him insane and walk on. A few years later, you become interested in time theory and build a time-machine. You travel into the past and meet a younger version of yourself in the same street. You identify yourself to him, but he thinks you insane and walks on. A few years later, *he* becomes interested in time theory, builds a time machine

. . . et cetera. Here again, the circle is closed, nobody in the story has any free will, and the events in it cannot happen.

It is all very well to say that they cannot happen, some people will object, but what if they *did* happen? The late science fiction writer Frederic Brown wrote an eerie short story called 'Experiment' to answer this question. Professor Johnson shows his colleagues his new time-machine. He has a brass cube in his hand. The time is six minutes to three. At exactly three o'clock, he tells his colleagues, he will place the cube on a platform inside the time-machine and send it five minutes into the past. 'Therefore,' he explains, 'the cube should, at five minutes before three, vanish from my hand and appear on the platform, five minutes before I place it there.'

At five minutes to three, the cube duly vanishes from his hand and appears on the platform, having been sent back five minutes in time by his future action of placing it there.

But one of his colleagues objects: what happens if at three o'clock he decides to change his mind and not place the cube on the platform?

Professor Johnson agrees that this would be an interesting experiment, and does nothing at three o'clock. There is no violation of causality. Instead, the whole Universe, including Professor Johnson, his colleagues and the time-machine, quietly disappears.

For a simpler view, take the classic short story by Sam Mines which he called 'Find the Sculptor'. A man builds a time-machine and travels five centuries into the future. There, he finds a statue commemorating himself as the first time-traveller. He puts the statue on board his time-machine and carries it back to the present. He then re-erects it in his own time. The catch here is obvious. The statue had to be erected in his own time so that it would be in the future when he went there to find it. But he had to go into the future to bring it back and re-erect it. The unanswerable question is: When was the statue made? All the people in these stories are pseudo-people, either with no possibility of existence or with their existence being brutally curtailed, like the unfortunate Professor Johnson.[5] They are pseudo-people because their authors have assumed the existence of one single dimension of time, one basic state of reality, in which an event can happen once and in one way only, no matter how many times the characters are made to revisit that event.

But why must we assume a single state of reality? Why should there not be an infinite number of parallel worlds? In these conditions, the circle would not close. Smith and Jones would have no difficulty in sending each other high-speed messages. There would be one world in which Jones received the message and another in which he did not. Both worlds, as well as an infinite number of others, would exist on different planes. This crazy-sounding idea has been seriously considered by scientists, and its paradoxes have been sharpened and elaborated by science fiction writers. It is usually called Everett's Metatheory (meaning something that is half theory, half speculation) after the Ph.D. thesis which the Princeton physicist Hugh Everett first published in 1957.[6] Everett proposed that from the birth of the Universe, reality has been branching into different states. There is no basic state of reality, as everyday experience would indicate, but instead an infinite number of 'relative states'. We live today in a relative state in which the Allied Powers won the Second World War. But according to Everett's Metatheory, there exists another parallel relative state in which the Nazis won. It is futile to assert that one of these states is 'true' and the other 'false'. Both are equally true, according to the Metatheory, and it depends simply on which one you happen to live in. 'You' might be a quite different personality in the relative state of Nazi victory, even assuming that 'you' were alive at all. But the matter does not end here, since the Metatheory predicts that reality branches in a different direction at every micro-instant of time. There is therefore a third relative state in which Hitler was shot dead in the Munich putsch of 1923, and as a result the Nazi party was split into feuding factions, and the war was never fought.[7] There are fourth, fifth and sixth relative states which grow progressively stranger and less recognizable. 'If there are infinite universes,' wrote Frederic Brown, 'then all possible combinations must exist. Somewhere, everything must be true. There are universes in which the states of existence are such that we would have no words to describe them or thoughts to imagine them.'*[8]

Whether or not these infinite relative states exist we have not the faintest idea. Everett appears to have intended his idea only as an interesting speculation. They represent what Brandon Carter might call a 'pathological' situation, but which does not violate causality, and so there is no need to invoke Everett's idea in support of what follows. Having now a clearer idea of causality and the conditions in which it would or would not be violated, let us see what Carter and others object to about instantaneous travel.

Smith, promoted to the status of astronaut from having been a mere sender of messages, enters a rotating black hole in the neighbourhood of the Earth and travels instantaneously to Planet X which is ten light-years away. His journey is truly instantaneous, by whatever clock its duration is measured. One moment he is near the Earth, and a micro-instant later he is in the neighbourhood of Planet X preparing to land. Now Jones is cruising around in another spaceship at a point slightly nearer to Planet X than to Earth. He has on board an amazingly powerful telescope through which he can see what is happening both on Planet X and on Earth. He sees Smith arrive on Planet X, and a moment *afterwards* he sees him depart from the Earth. There is

*Isaac Asimov's novel *The End of Eternity* is the best fictitious treatment of the Metatheory I have read. A group of bureaucrats survey human existence from outside time and make continual changes in 'reality' for humanity's good. Needless to say, their constant meddling interference proves catastrophic for the human race, since the bureaucrats, being the kind of people they are, always choose the safest possible reality. Any course of events liable to prove dangerous or adventurous is always excluded.

nothing surprising about this. Since Jones is slightly further back from Earth than he is from Planet X, the light signal from Earth takes longer to reach his telescope.

But suppose that Jones, with his equally amazing radio transmitter, decides to play a practical joke on Smith. He sends a message to Earth saying that war has just broken out on Planet X and that a trip there would be dangerous. He urges Smith to stay at home. This is where Carter and his colleagues suspect a violation of causality, for Smith's arrival on Planet X has already been observed. What happens if Smith heeds the warning and doesn't go? We seem to have reached the situation in which Smith's journey can take place only if it does not take place.

And yet there is a simple answer. Jones's frivolous message indeed reaches the Earth but it does not reach Smith. There is no violation of causality because it arrives several years too late. Jones has forgotten in his prankish enthusiasm that he is slightly more than five light-years from the Earth, and that a radio message can travel no faster than the speed of light.

No faster than light? This answer must sound like one of those 'shaggy dog stories' whose humour lies in the unexpected rejection of a concept that was quite acceptable in the previous sentence. We have been talking for many pages now about the feasibility of instantaneous travel, and now we suddenly overrule an apparently well-argued objection to this concept with the plea that a message cannot travel faster than light.

Very well, let us bow to this criticism and say instead that Jones has a machine which both receives and transmits messages instantaneously. He learns of events both on Earth and on Planet X the instant that they happen, and every message that he sends is received the next instant on either planet. Again there is no problem. He *can*, in this case, dissuade Smith from making his journey. Smith will receive the message in time to remove his luggage from the spaceship and cancel his ticket. It is perfectly proper for him to do this, since Jones never saw him arrive on Planet X. Nobody saw him arrive on

Planet X; which is not surprising, since he never arrived there. Since the messages reaching Jones were instantaneous, he only saw events happen in the order that they actually happened. There was no violation of causality because the circle did not close. One major objection to instantaneous travel is now answered: we can see a man end his journey before he has begun it, but we cannot, in the light of this information, communicate with him in time to prevent his departure. And we cannot use instantaneous communication to detect a violation of causality because if this mode of communication is used there is no violation of causality. Events can only be observed to happen in their correct order.

Events happening in their correct order? Some students of Einstein's Special Theory insist that events have no correct order, and that there is no way to prove beyond the possibility of doubt that one event preceded another! It is not a matter of being unable to establish the absolute truth, they say; there is no absolute truth. There is no such thing, they believe, as provable simultaneity of events. This is a vital point, since if they were right and it was impossible to prove simultaneity, an instantaneous journey would be inconceivable. The time of arrival, as defined by the phrase, 'A microinstant after departure', would vary according to the viewpoints of different observers, and there would be no absolute and agreed time of arrival.

Some time ago, two jet aircraft were in mid-air collision over Mexico. It proved exceptionally difficult to discover the cause of the crash. At length one investigator exclaimed in exasperation: 'The only thing we are sure of at this stage is that the two aircraft ceased functioning simultaneously.' This indeed would be the only thing that the rival schools of relativists would be sure of. All would agree that the disintegration of the two aircraft was truly simultaneous, and that there is no question of the first having struck the second before the second struck the first. But when events are separated by great distances, there is surprisingly no unanimity about which of them happened first. The mathematician Martin

Gardner, deputy editor of the *Scientific American*, once constructed an ingenious tale which may be called the paradox of the astronomer's broken leg.[9] It goes like this: an astronomer somewhere on Earth falls off his telescope and breaks his leg. The experience is painful, and for that reason very memorable, and the astronomer makes in his diary a careful note of the date and hour of the accident. An astronaut friend of his is during this time returning from a visit to that familiar and convenient outpost, Planet X. The astronaut here is not travelling instantaneously, but is moving at a speed slightly slower than light. He lands on Earth, greets his old friend, and they talk as follows:

> 'I've had a terrible time,' says the astronomer. 'I fell off the telescope and broke my leg and it never mended properly. It still hurts.'
>
> 'Poor fellow,' says the astronaut. 'When did that happen?'
>
> 'Five years ago,' says the astronomer with a grimace.
>
> 'Five years ago? That can't be right. I hadn't left Earth for Planet X then, and your leg was O.K. at that time.'
>
> 'What nonsense! You weren't here five years ago. You set off on your journey many years before that, and here is my recording clock to prove it.'
>
> 'And here is *my* recording clock that proves that I left for my journey considerably *less* than five years ago.'

Both clocks are working correctly. What has happened is that the astronaut's clock had slowed down when he was travelling at close to the speed of light, while the astronomer's clock had registered a constant Earth-time. By an examination of clocks alone, there is *no way* of establishing exactly how many years ago the astronomer broke his leg. No two events—in this case breaking the leg and the recording of the accident—however simultaneously they may seem to have happened, may be proved by looking at clocks to have been truly simultaneous. But this is only true if the investigation is restricted to the recordings of clocks. The astronomer wrote in his diary the date and the hour of his accident, and with this figure there can be no argument. The astronaut can easily confirm from his own records that he was flying through deep space on that date. He belatedly remembers Einstein's principle of the slowing of time, and the mystery is cleared up.

Here is one of the most widespread misunderstandings of the Special Theory of Relativity. It is wrongly supposed that the slowing down of time inside a fast-moving vehicle somehow does away with the simultaneity of events. All sorts of people who should know better are constantly either making this error or criticizing others for not making it. They appear to have forgotten that Einstein himself firmly established the principle of simultaneity in a series of well-publicized lectures in 1921.[10] A little reflection should nevertheless convince us that they are wrong. Imagine a section of the Galaxy (perhaps one of the globular clusters with their tens of thousands of closely-packed stars) where there might be intelligent life, high technology and a continuous traffic of fast-moving spaceships.* Some of the inhabitants stay at home on their various planets, while others are engaged on interstellar journeys at speeds close to that of light. At a certain instant, a person on the surface of a planet ties his shoelace. At that very same instant, another person, in a spaceship moving close to the speed of light, starts to smoke a cigar. Now from the viewpoint of the man tying his shoe on the planet's surface, the other man's act of lighting his cigar might take several days. The man in the spaceship puffs away at his cigar and finally stubs it out after about thirty of his minutes. Yet while he was smoking, the man who was tying his shoe may have lived through an entire month. (I say, 'may have' since I haven't bothered to specify exactly how fast the spaceship is moving.) Meanwhile another man, on another spaceship which is moving not quite so fast,

*I am not of course suggesting that there is any life, intelligent or otherwise, in any particular globular cluster or any other given point in the Galaxy. This is purely a thought experiment.

eats his breakfast while the shoe-tier lives through a fortnight. And so on, depending on how close the various spaceships are to the speed of light. There is an infinite number of 'local times', of local speeds of time, which depend in turn on the speed through the Universe of the clocks that measure them. But amidst all these local speeds of time, there is an absolute and provable simultaneity of specific events. There was a provably simultaneous instant when one man was tying his shoe with relative rapidity while the other was at some point in the age-long process of lighting his cigar. This instant was infinitesimally small, like a point on a geometric line. But still its existence is beyond dispute.

And so we have established simultaneity, which is essential for instantaneous travel. There remains a more subtle problem of a different kind involving causality, which I will call the paradox of the infallible fortune-teller. Some people believe that the interior of the black hole itself presents an inherent violation. An everyday version of this paradox is already familiar to many New Yorkers. According to a corny old story, a man falls from the top of the Empire State Building. As he passes the twentieth story on his way to the ground, he is heard to mutter: 'Everything's fine so far.' Now is there a violation of causality if a fortune-teller with supernatural powers of infallible prediction leans out of a window at the twentieth floor and calls out: 'No, sir. Everything is not fine. Your death within a matter of seconds is inevitable'?

There is a violation here, but it is so trivial that it would only matter in one such case in a hundred. If a hundred men fell off the Empire State Building at different times, it is conceivable that one of them might survive, and that his rescue might be effected *after* he had passed the infallible fortune-teller on the twentieth floor. A huge hay-cart, piled with thick, soft hay might appear on the street below him in time for him to fall into it. (It would have to be a miraculously large hay-cart to cushion his tremendous speed, but almost any improbable detail, including an infallible fortune-teller, is permissible in a thought experiment!) Or the Fire Department, noting that an unreasonably large number of people were falling off the Empire State Building that day, might have rigged up a system of nets of great elasticity that would shoot out from somewhere around the tenth floor and catch the people as they fell. In this minority of cases, causality would have been violated because the fortune-teller cannot be both infallible and wrong. If she is wrong she cannot be infallible, and if she is infallible she cannot be wrong. And so there is a clear violation.

Some physicists worry that the same paradox might arise as a spaceship vanished through the outer event horizon of a rotating black hole. As soon as it passes the outer event horizon (with its automatic pilot pointing it correctly towards the inner event horizon as proposed by Figure 4) (see p. 46) its future is as inevitable as the fate of the man falling off the skyscraper was not. If the chances of imminent death of the falling New Yorker were only 99 per cent, the chances of the ship proceeding down the black hole on its ordained course are truly 100 per cent. There will be no cosmic equivalents here of nets or hay-carts. And so it would be possible, in the mad world of thought-experiments, for the spaceship to meet an infallible fortune-teller after it has passed the outer event horizon. She could then prophesy: 'You are destined to pass through this black hole into another region of space, and this destiny is unalterable.' There would be no chance whatsoever of her prediction being wrong. One physicist, who knew more about causality than he did about black holes, asked me with a frown: 'What happens if the captain of the spaceship is determined to prove the fortune-teller wrong, and swings his ship around and heads back out of the black hole by the same route that he came in?' The answer is that he cannot. Nothing can repass the event horizons of a black hole. As I explained in Chapter 1, a black hole becomes black when nothing that flows into it, not even light, can flow back. And so the fortune-teller in this case is truly infallible, and causality is not violated.

There finally remains the paradox of the spaceship appearing to be in two places at once. If we are watching its journey from afar, we do not see it actually vanish past the event horizon. Instead, we see the spaceship as it was just before it vanished; and we continue to see it in this position for many years! It appears to us to have become trapped in orbit outside the black hole, unable equally to penetrate into it or to escape from its neighbourhood. It is plain what has happened. The light image of the spaceship, the very photon light particles that made up that image, have themselves become trapped in orbit round the black hole on the very surface of the event horizon. They did not approach the event horizon near enough to vanish, and since this image, being a mere illusion, cannot escape from orbit, we thus observe, or imagine that we observe, an everlasting and purposeless voyage: the vision of a cosmic rocketing Dutchman.* How strange it would nevertheless be for the distant observer if, while watching this sight, he received a message from the captain of the spaceship reporting his safe arrival on Planet X. Only a trained theorist would have the knowledge not to say: 'But you haven't arrived on Planet X at all. I can see you trapped in the orbit of the black hole!' A man may appear to be a thousand places at once in relativity theory, but he must truly be at one place at one time.

This odyssey through the paradoxes of space and time must now yield to a discussion of the practical engineering problems of using black holes for Galactic transport. It was necessary to show that instantaneous travel through the Galaxy would violate no fundamental law. Now, in the hope that I have shown it, we must face the

*The image would in fact be gradually broken up by the violent perturbations of fields around the black hole, and would become progressively more blurred and faint before eventually vanishing altogether.

question which may have puzzled some readers: the nearest known black hole is 6,000 light-years away from us in the constellation of Cygnus. What are our descendants going to do about those 6,000 light-years?

Part Two

THE IRON SUN

Had I been present at the Creation, I would have given some useful hints for thc better ordering of things.

ALFONSO THE WISE, KING OF LÉON AND CASTILE (1242–84)

5
The Arm of Orion

One of the finest sights on Earth is a splendid sunset. This fact is well-known to advertising men, who often find that the easiest way to sell us a holiday on a tropical island is with pictures of a waving palm tree against the backdrop of a darkening horizon streaked with fantastic cloud patterns of purple and crimson and gold. Perhaps, if sunsets did not fade so quickly, one could stare at them in hour-long reverie, drawing mental pictures from their random shapes, as we do when we gaze into the flaming embers of a fire. As a poet once described these phantom images:

> The fair, frail palaces,
> The fading alps and archipelagoes,
> And great cloud-continents of sunset seas.[1]

And yet spectacular sunsets are caused by a most unromantic substance in the Earth's atmosphere, namely dust. This is natural dust, not industrial pollution which, so far at least, is only significant in localized regions. From hearing the complaints of environmentalists, one might imagine that before the industrial age the air of our planet had a translucent, almost winelike purity. Nothing could be less true. The great haze that lies over the Amazon is probably as old as the river itself. Volcano eruptions have periodically blackened the skies for hundreds of miles around them. For every particle of dust which escapes from a factory chimney today, thousands of millions more come either from volcanoes, are swept up by winds from the deserts, or arrive as meteoric dust from space. Very few people have actually been struck on the head by meteoric stones, or even seen or heard them fall, but it has nevertheless been estimated that about 1 million tons of dust and fragments of rock from space land on the Earth each year.

Those wondering what all this has to do with black holes, and the problem of travelling 6,000 light-years to the nearest known black hole, must be patient a little longer. Dust, second only to stars, forms the main aggregate of mass in our Galaxy, and we shall presently see how these immense quantities of dust that lie in interstellar space can be bent to our purposes. For deep space is very 'dirty'. In the words of one astronomer, 'nature is a messy housekeeper'. Space is millions of times more transparent than the Earth's atmosphere, but as soon as we try to peer through telescopes into its immense distances, we find often that our vision is wholly obscured by clouds of cosmic dust.[2]

The broad swathe of innumerable suns that make up the Milky Way (another name for our Galaxy) appears to us jagged and irregular. We see in it nothing of that orderliness and curved symmetry which thousands of millions of years of Galactic evolution and the developed orbits of stars around one another might lead us to expect. Everywhere there are huge lanes of blackness and dark blotches and spots. There is, for example, a strange dark shape of utter blackness in the constellation of Orion which resembles the head of a monstrous horse. Known as the 'Horsehead Nebula', it is actually several hundred light-years from ear to mouth, and it is nothing more than a random configuration of dust surrounded by the white aura of thousands of bright stars. The 'Coal Sack' is another ominous-seeming pit of darkness, this time in the constellation of Crux, the Cross. Imagine a great globule of blackness appearing against a background of brilliant colour. The Coal Sack looks like some evil cancerous growth amidst the purity of the heavens. One can imagine the terrible myths which the ancients might have constructed around the Horsehead and the Coal Sack if they had had telescopes to peer into the Galaxy and see them.

It might be supposed, from the deep black appearance of these objects, that an astronaut who plunged into one of them would find himself enveloped in choking clouds of soot and that his visibility would be reduced to zero. This idea is completely wrong. Dense though the dust-clouds may be in comparison with some of the surrounding star-fields, they are none the less millions of times less dense than terrestrial air at the peak of a high mountain in perfect weather. Their density is unbelievably low. Two flies buzzing around in a large room are closer together than the particles that comprise these dust-clouds.[3] And the particles themselves are much smaller than flies; their diameters are no more than a few thousandths of an inch.

A few statistics are essential here. Interstellar space has an average density of one hydrogen electron per cubic centimetre. Hydrogen is best suited for this illustration, since the mass of the Universe, including both the stars and the interstellar clouds, is mostly hydrogen. Now the mass of the electron of a hydrogen atom is 10^{-27} grams —that is, a decimal point followed by 26 zeros and then a 1.* We can say therefore that the average density of matter in interstellar space is 10^{-27} grams per cubic centimetre. But we are not concerned here with average densities. We are concerned with the densities of the dust-clouds, some of which are 100 million times denser than the average density of space outside them.[4] This density is therefore $6 \cdot 3 \times 10^{-21}$ pounds per cubic inch (about 10^{-19} grams per cubic centimetre), which is still almost a perfect vacuum compared with the density of the air that we breathe, but it is this differential factor of 100

*This statistic is so well-known to astronomers that it would seem almost criminal to call it $3 \cdot 6 \times 10^{-29}$ pounds per cubic inch. I decided to write the text of this book (although not the Notes and Appendices) in imperial measurements, but there is something sacred about the mass of the hydrogen electron.

million that accounts for the blackness of the Coal Sack and the Horsehead.* Now this difference in density is important, for this is the phenomenon of nature which can solve the problem that will face our descendants: What to do about those 6,000 light-years that lie between them and the nearest known black hole.

We must nevertheless continue for a while with our mysterious discussion of dust-clouds. I am not being intentionally mysterious; but the concept of the 'Iron Sun' will only be intelligible to someone who has some general grasp of the density and the geography of these clouds.

The Galaxy of stars which surrounds our Sun and its planets is a galaxy of 'spiral' type. Our Galaxy contains about 180,000 million suns. Some idea of the size of our Universe may come from the estimate that there exist about 80,000 million *other galaxies*. This is a very loose estimate, which may be as much as 20 per cent too few or too many. The galaxies are so numerous that there is simply no way of counting them accurately. (I have also a suspicion that any project that involved the meticulous counting of galaxies, by, say, numbering them to the nearest million, would be considered unscientific and the waste of a telescope's valuable time, because it would be unlikely to reveal any general cosmic principles.) Apart from exploding or highly peculiar galaxies, which need not concern us here, they come in three general types: the ellipticals, the spirals and the barred spirals. An elliptical galaxy, as its name suggests, is shaped much like an egg except that both ends are of equal size. A barred spiral resembles two scythes whose handles are fastened together so that the two blades point in opposite directions. Our Galaxy is neither of these types. It is an ordinary spiral.

Anyone who has ever seen even the most modest

*As has already been said, all these different densities must sound confusing, so in the Glossary, under 'density', I give a list of various conditions of density to be found on Earth and in different parts of the Universe.

display of fireworks will be familiar with a particularly beautiful type known as a Catherine wheel. This firework is constructed in the form of a small cross, with little rockets fixed into each of the four points. Through a large hole in the centre, it is nailed loosely to a wooden post so that it can rotate freely. A single fuse ignites all four rockets simultaneously. Powered by these rockets, the whole firework begins to rotate. The fine spectacle at night is produced because the small centre, spinning loosely round the nail, gives the illusion that it is rotating much faster than the extended outside points. This illusory *differential rotation* makes the fiery rocket trails form beautiful spirals round the fast-spinning centre.

The mechanism of a spiral galaxy such as our own, although on a vastly greater scale, is similar to that of a Catherine wheel. The central hub, or nucleus, is equivalent to the parts around the nail around which three or four extended spiral arms are rotating. A complete revolution of one of the arms around the Galactic centre, instead of being completed within a second like the spirals of a Catherine wheel, takes about 200 million years.*

How can we be sure that this hypothesis is correct? The centre of our Galaxy, somewhere in the constellation of Sagittarius, is obscured from us by clouds of interstellar dust; the dust-clouds, being about 100 million times denser than the surrounding space, obscure everything behind them from sight. And so we cannot see whether the centre of the Galaxy consists of a large group of densely packed stars, as some theorists have suggested, or, as others believe, of a gigantic black hole which would devour stars almost as soon as they are formed.[5] But this difficulty has been partly resolved because we can peer much more closely at the centres of

*Since the age of the Galaxy is about 10,000 million years (twice the age of the Sun), the Sun and the Earth have probably revolved around the Galaxy more than twenty times. This rotation period of 200 million years is known as one 'Solar year' or 'Galactic year'.

other spiral galaxies. The fact that they are millions of times further away from us than the centre of our Galaxy makes no difference; there is little obscuring dust in intergalactic space, and so we see them clearly.

One of the finest astronomy books ever published is the *Hubble Atlas of Galaxies,* edited by the California astronomer Allan Sandage and published in 1961. Although usually unavailable in ordinary large bookshops, it is essential reading for anyone who wishes to learn anything about the nature of galaxies. Indeed, I sometimes believe that one can learn more in half an hour about the large-scale physical structure of the Universe by glancing at the photographs in this book than in half a year of reading a million words of print.[6] After a brief and lucid introduction by Sandage, the rest of the book consists of about sixty high-quality photographs of those galaxies nearest to us, taken during the last forty years through the powerful Californian optical telescopes at Mount Wilson and Mount Palomar. With spiral galaxies in particular, there is an almost perfect clarity. And since so many spiral galaxies are photographed in the book, appearing to us at so many different angles from our vision, we soon acquire precise ideas about the shape and nature of our own Galaxy.

We cannot claim to know a man very well if all we have ever seen of him is a single photograph. But if we can see twenty or thirty photographs of the man, taken from many angles, showing him dressed in a variety of costumes, and at different times in his life, from early youth to old age, we can say that our knowledge of the man, if not profound, is certainly substantial. So it is with the 20,000 million or so spiral galaxies, of which a representative sample are so clearly pictured in Allan Sandage's *Atlas*. The first thing that strikes us, after making allowance for the indistinctness that begins to blur the details at distances beyond about 20 million light-years, is that spiral galaxies all resemble one another very closely. All have similar numbers of arms spiralling away from the nucleus and therefore look strikingly similar. All the spiral arms of these galaxies have roughly the same proportions. In all are shown plainly the effects of differential rotation which we can see in a spinning Catherine wheel. Obviously therefore these pictures of the Universe outside our Galaxy are of great importance to the astronomer who is mainly interested in the state of affairs inside it. For despite the obstruction of the dust-clouds, we can make out just enough detail of our own spiral Galaxy to realize that its shape is identical to that of thousands of millions of other such galaxies in the Universe, whose details are so clearly shown in the many pictures in the *Atlas* as to make them as plain and as familiar as a road map.[7]

The intensity of investigation into various topics of science seems to vary in cycles. For many years, until about 1974, few people seemed very interested in the composition of those spiral arms of dense matter, which sweep through the Solar System like giant torchbeams, black from dust and brilliant from starlight, at intervals of about 100 million years. But the subject has now risen again in a most unexpected way. For several years there have been claims that the world's climate, at least as far as the northern hemisphere is concerned, is becoming colder and that a new Ice Age may be beginning. The average temperature is alleged to have fallen by a degree in the last thirty years; the Arctic glaciers are said to be advancing southwards, and the annual growing season in several northern countries has apparently shortened by about a fortnight.

Now many of these findings have been disputed, and two exceptionally hot summers (in 1975 and 1976) have, for the time being at least, allayed the fears of those anticipating an imminent new Ice Age.[8] But the whole controversy has provoked a much larger question: What conditions could produce an Ice Age? The Sun, obviously, would be the main culprit. If its radiation was interfered with for any reason, it would become colder on Earth. But the long-range climatologists and their astronomical friends would not be doing their jobs if they were merely to observe the Sun and report any changes in its behaviour. For we can easily imagine a case in which the Sun is radiating normally but *something* has come between the Earth and the Sun, and is interfering with that radiation.

I do not mean that any unusual object *is* interfering with the sunlight, but simply that one day it *might*. What manner of object or objects could these be? Professor W. H. McCrea, of Sussex University, gave in 1975 a comprehensive answer. He pointed out that the Solar System has recently entered one of the spiral arms of the Galaxy, known to galactic astronomers as the Arm of Orion.* [9]

The Orion Arm has little or nothing to do with Orion the Hunter, and his famous 'belt' and 'sword', whose delineaments the ancients fancied they saw in the star-patterns of their northern winter skies. The arm is so-called because we see it, disfiguring part of the Galaxy, in the Orion constellation. Professor McCrea's hypothesis is that the Ice Ages which have occurred periodically during the Earth's long history might have been caused by the Solar System's passage through one of the Galactic spiral arms. Passage through a spiral arm could fill the volume of space between the Sun and the Earth about 100 million times more densely with dust particles than it normally is. Some of this extra dust would fall into the Sun and heat up its surface layers, by kinetic energy, so that the solar radiation reaching the Earth would therefore become hotter. Paradoxically, this greater heat could set off an Ice Age, by increasing the Earth's cloud cover and in turn the level of rain. The clouds would deflect away much more of the Sun's radiation than they

*When I say 'recently entered', this does not of course mean within the last few years or so. We may have entered the Arm of Orion perhaps 1,000 years ago, or at some date even more remote. It would have been difficult for our ancestors to record such an event. Until the 1920s, when Edwin T. Hubble discovered that our Universe consists of thousands of millions of separate galaxies, people knew little or nothing of galactic structure.

normally do. This would have the effect of lowering the average temperature on the Earth's surface by one or two degrees. This lowering of temperature, within the space of about 1,000 years, would enable the glaciers to expand, and eventually cover Europe with sheets of ice as far south as Gibraltar, and North America as far south as St Louis.

McCrea's theory is one of several put forward to explain why there were so many such massive invasions of ice during prehistory. Yet several geophysicists feel that there is another explanation which has nothing to do with any interference of the Sun's radiation. Their theory should be briefly explained. The name North Pole is nothing more than a convenient geographical expression. Our compasses do not point towards the North Pole. They point instead to the Magnetic North Pole, in a direction often called 'true north'. This Magnetic North Pole, to which our magnetic compasses point, is now situated near Prince of Wales Island, in northern Canada, about 1,300 miles from the 'real' North Pole. This point continuously changes its position. According to the evidence of fossil crystals in the Earth's crust, 350 million years ago the Magnetic North Pole was near Japan; 500 million years ago it was near Hawaii. Only in the last 10 million years or so has it been in the Arctic Ocean. The icy polar regions of the Earth move with the Magnetic North Pole. The world's geographic regions have thus shifted from polar zones, to temperate regions, and then back again, on countless occasions. This theory agrees remarkably well with the evidence of fossils, particularly with those remains of tropical animals and flora which have been found in the ice-ridden Anarctic.[10]

Another difficulty with McCrea's tentative explanation of Ice Ages is that the volume of space between the Earth and the Sun appears to be much too small even for a 100 million-fold increase in its average density to have any significant effect. This volume is a sphere with a radius of 93 million miles, the average distance between the two bodies. This is a tiny distance on the Galactic scale. It is a mere sixty-thousandth of a light-year, and about 2,000 millionths of the length of the curved Orion Arm. This little sphere of 93 million mile radius has always contained a large quantity of matter, from fragments of burnt-out comets to pieces of stray asteroids, which sometimes crash into the Moon, where huge craters mark their impacts. Setting aside these quantities of wandering debris, which are part of the normal structure of the Solar System, we find that the total mass of the scattered particles which have come between the Earth and Sun as a result of our passage through a spiral arm is only 10^{15} tons, which is about one six-millionth of the total mass of the Earth, or one half of a trillionth of the total mass of the Sun.* It must be seriously doubted whether such a small mass of dust particles is sufficient to interfere with the warmth that we receive from the Sun.[11] It is true that the periods of glaciation in prehistory appear roughly to coincide with the times of our previous passages through the Galaxy's spiral arms, so far as the times of those passages can be calculated. But they concur equally well with the known movements of the Magnetic North Pole which, from the evidence of fossils, can be much more reliably documented.

But if the Orion Arm is unlikely to bring about an Ice Age because of its insufficient density, it is none the less extremely dense on the Galactic scale. It should now be clear that we on this planet have little to fear from its intrusion into our lifeline from the Sun, because our distance to the Sun is so small. But imagine the scale of the Arm's density over the distance of a light-year, 6 trillion miles, a distance 60,000 times greater than the 93 million miles that separate the Earth and Sun. Every second we move more deeply into the Orion Arm, and

*By a half of a trillionth (a somewhat clumsy phrase), I mean one half of a millionth of a millionth. This number is written mathematically as 5×10^{-13}.

every second there is an increase in the quantity of dust and interstellar hydrogen that lies in a spherical region with a radius of one light-year round the Sun. It will be about 50,000 years before we emerge from this dust arm, and we shall be a poor species if, long before that period has elapsed, we cannot exploit this immense quantity of dust for our technological purposes.

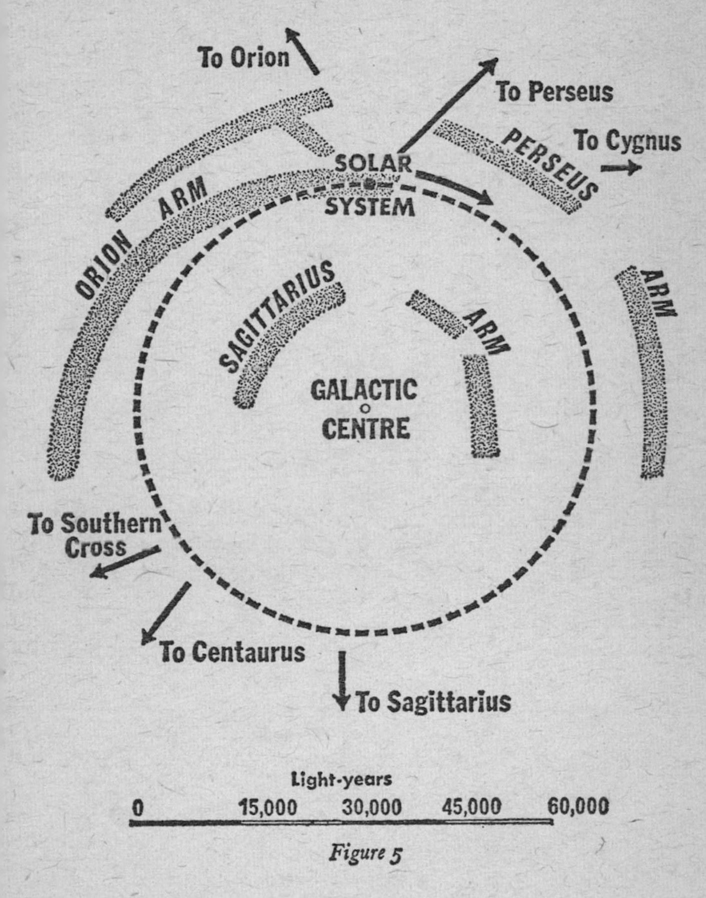

Figure 5

What is the 'geography' of this galactic dust? Many students of astronomy are familiar with diagrams of the Galaxy that look like two saucers inverted upon one another. But if one were able to look 'down' upon the Galaxy from some great distance 'above' it, one would see something like the shape which is drawn roughly in Figure 5. The spiral dust-arms are plainly shown. The rotation of the Solar System around the Galactic Centre is clockwise, as the arrow in the diagram indicates. The directions of some of the constellations are shown. The Solar System takes about 200 million years to make a complete circuit of the Galaxy, and so several thousand years more will elapse before it clears the Orion Arm.

This dust is more than just the debris of a galaxy. Together with hydrogen, it is the material from which new stars are formed. Six thousand million years from now, our Galaxy will look much as it does today. But a close inspection will reveal one great difference: most of the stars that we know today, including the Sun, will have vanished, and the Galaxy will consist of new stars formed from the dust and hydrogen that now make up the spiral arms. It is interesting that this process is already happening: a star many light-years from the Sun and inside the Orion Arm has recently been observed in the actual process of being formed.[12]

What does galactic 'dust' consist of? Much of it was apparently created in turn by a still older generation of stars which were shining long before the Sun was formed. Many of these ancient stars blew themselves up in massive 'supernova' explosions, scattering their fragments over vast distances. Among these fragments are great quantities of nickel and iron that came from the cores of these exploding stars. And of all those million tons of meteoric dust which fall onto the Earth from space today, by far the greatest part is of nickel and iron that originated in the Orion Arm, the materials which long ago played a part in the birth of our own Sun.

It is an educational experience to visit the minerals section of the Natural History Museum in London. On the first floor, there is a long corridor devoted to terrestrial minerals. They are brilliant in their colours and a delight to look at. There are all manner of green, yellow and purple rocks and crystals, distorted into fantastic shapes as they exemplify many complex forms of beautiful oxides and chrondites and other formations. At the end of the corridor is a darkened chamber heralded by the simple notice, 'Meteorites'. Here there are no brilliant colours. All the substances here have a more sombre appearance, their colours restricted to grey and brown and black, for nickel and iron and their compounds comprise most of the meteorites on display. Here is the very stuff of the stars, cooled but otherwise unchanged since it was belched forth from the furnace of exploding suns. One of these compounds is the ferric oxide known as hematite. Astronomers have found that interstellar dust is richly mixed with hematite, and its presence in meteorites confirms their observations.[13] The prize exhibit in this part of the museum is in the furthest corner. It is a lump of nickel and iron, about a yard high and a yard across. Known as the 'Cranbourne meteorite', it landed near a village of that name in Victoria, Australia, and was discovered there half-buried in 1854. But here is the astonishing statistic: this little lump of rock the size of a small armchair weighs no less than *three and a half tons*.

For an atom of iron weighs fifty-six times more than an atom of hydrogen. Iron is by no means the heaviest element in the periodic table, but there are few heavier among the common elements. It is also the commonest metal in the Universe. The hydrogen in the Arm of Orion is richly mixed both with nickel and with iron, as we have seen. This probably means that in every cubic centimetre of the dust in the Orion Arm there is at least one atom of nickel and one atom of iron.

In short, in a sphere with a radius of one light-year round the Sun, there is a huge quantity of matter. This matter is widely diffused. But it is possible to estimate roughly its total aggregate of mass. The figure amounts to something between 100 and 1,000 times the mass of the Sun. This volume of material is many times more than our descendants will need if they wish to find the raw materials with which to construct a black hole.[14]

Substances consisting of nickel and iron have another useful property; in certain circumstances they can be moved across great distances by the field of a magnet.

6
The Astromagnets

A remarkable episode took place in the James Bond spy film *You Only Live Twice*. Bond's car was being chased through the streets of Tokyo by a carful of gangsters who were firing at him with automatic rifles. Bond radioed to the local secret service for help. It arrived quickly. A helicopter appeared and flew low above the gangsters' car. Suspended below it on a cable was a huge electro-magnet. The helicopter pilot threw a switch, and the gangsters found themselves in a strong magnetic field. Their car leaped vertically from the road, and its roof struck the bottom of the magnet with a great clang. The helicopter then carried the car, with its four bewildered gangsters, to an altitude of 2,000 feet. It crossed a near-by coastline and hovered over deep water. The pilot then switched off the magnetic field. The car was at once released, whereupon it fell into the sea, killing all four men.

This scene, although exaggerated for the purposes of fiction, gave a good indication of the power of modern industrial electro-magnets. Such magnets are used today for many tasks. They generate electricity in power stations, and I have watched them picking up great heaps of scrap-iron in steel factories. On a smaller scale, magnetic units are used in many everyday devices, from domestic frigidaires and vacuum cleaners to telephone receivers. Several million tons of metal are processed each year solely for use as magnetic materials. The magnetic force, as its suitability for so many uses might suggest, is one of the strongest of the fundamental forces of the Universe. Somewhat more complicated in its nature than gravitation, since it both attracts and repels, it is in fact many trillions of times stronger.* We can get some idea of its strength from a child's toy which performs a fascinating experiment with a magnet and iron filings. One can buy in almost any toyshop the equipment for this experiment. It consists of a small magnet and a bottle of iron particles, so finely-shaven that they look like dust. One scatters these iron filings at random on a sheet of thin paper, which is held flat and taut. The magnet is then moved slowly just beneath the paper. The iron filings at once assume a strange and beautiful symmetrical pattern. They can be made to delineate precisely the complex shape of a figure-of-eight magnetic field as accurately and as graphically as most textbook illustrations.

We speak so far only of iron. Iron, nickel, cobalt and their alloys are the metals most strongly attracted by magnets, and are called *ferromagnetic*. They are attracted by a magnet when in their normal cold states. No other substance is noticeably magnetic at room temperature.[1] A character in the Gilbert and Sullivan opera *Patience* laments this state of affairs in a sad song about a still-born love affair between a magnet and a silver churn:

> A magnet hung in a hardware shop,
> And all around was a loving crop
> of scissors and needles, nails and knives,
> Offering love for all their lives;

*Gravitation, mighty though it may appear, is the weakest of the three known fundamental forces. Magnetism, usually called the electro-magnetic force, is 10^{36} times stronger than gravitation. The most powerful of all is the nuclear force, which binds together the neutrons and protons in the nucleus of an atom. The nuclear force is about 100 times stronger than the electro-magnetic force or 10^{38} times stronger, than the gravitational force.

But for iron the magnet felt no whim,
Though he charmed iron, it charmed not him;
From needles and nails and knives he'd turn,
For he'd set his love on a silver churn!

But this magnetic,
Peripatetic
Lover he lived to learn,
By no endeavour
Can magnet ever
attract a Silver Churn![2]

Yet this statement is only true if the substance facing the magnet is at an ordinary room temperature. If sufficiently heated, *all* substances become magnetic.

There is a fourth state of matter which can exist beyond the three familiar states of solid, liquid and gas. It is called *plasma*. When any substance is heated to beyond about 15,000° Fahrenheit, it becomes plasma. Plasma is matter in which the electrons have been stripped away from their atoms. Without their encompassing electrons, atoms cease to be atoms. Instead, they become magnetically-energetic sub-atomic particles. This is the technical way of saying it. In plain language, any substance which is heated to various critical temperatures (the maximum for all substances being about 15,-000°) becomes as attractive to a magnet as if it were nickel or iron. The matter which they comprise has been *ionized*, or converted into magnetic plasma.

A large proportion of the physical matter in the Universe consists of plasma. The stars themselves, being extremely hot, consist largely of hydrogen and helium plasma. Such extreme conditions occur rarely on Earth, and plasma has been difficult until recently to isolate and study. It exists in the heart of lightning bolts and in that spectacular collision between the atmosphere and electrically charged particles from the Sun, which we call the Aurora Borealis, or 'Northern Lights'.*

*The brilliant Northern Lights, with their blaze of many colours and the ominous crackling noises which accompany their ap-

Plasma is produced artificially in the doughnut-shaped rings of experimental thermonuclear fusion reactors, where a particular form of hydrogen must be heated to a temperature of 200 million degrees Fahrenheit if it is to undergo fusion reactions. Fusion scientists, trying to imitate the power of the Sun in order to give us almost limitless electricity on Earth, experience great difficulty in trying to keep the plasma from hitting the walls of the ring-shaped container, where it immediately cools down. For only plasma is suitable for a fusion reaction; matter in any of the other three states would smash the curved walls of the reactor's container long before it reached the critical temperature. The plasma is kept away from the container walls by the only force that can control it, the force of a magnetic field. But this magnetic control has proved to be extremely difficult to attain in the small space of the container. No sooner is the plasma deflected in one place than it hits the wall in another. This collision lowers its temperature to below the critical level of 200 million degrees Fahrenheit, and the chance of achieving a reaction is gone. To deflect the plasma at every point, the scientists experiment with intricate arrangements of differently placed magnetic fields, which they appropriately call 'magnetic mirrors'. Fashioning their containers into ever more ingenious geometric designs, their struggle to achieve fusion goes on, and many of them believe that success may be imminent.[3]

I come now to a proposal which may, at a first consideration, be regarded as raving insanity. I am going to suggest that magnetic fields could be created in the dense Orion Arm of the Galaxy with the purpose of concentrating a sufficient quantity of matter at one point and actually constructing a black hole. The creation of a black hole, of course, requires far more energy than any

---

pearance, were long an object of terror to the ancients. In the twelfth-century *Saga of King Olaf*, they were identified with the war-god Thor and his flashing eyes and rattling chariot wheels. The lights over the southern magnetic pole, the Aurora Australis, are equally spectacular.

human society could hope to generate unless it was aided by nature. But stealing energy from nature has been a common practice of inventors since the seventeenth century, when Francis Bacon, the great philosophical forerunner of modern technology, made his famous rule: *Natura non vincitur nisi parendo.*[4] This may be freely translated: 'Never do any work if you can get the Universe to do it for you.' * The rule is directly applicable to this case. It has already been seen that iron and nickel, in their usual states, are magnetic. If hydrogen, the interstellar hydrogen that exists in relatively high densities inside the Orion Arm, could be heated up until it becomes plasma, it also would respond to a magnetic field.

How could our descendants possibly hope to heat up this interstellar hydrogen? The answer is very simple: they could do it by shining very intense beams of light, known as laser beams, directly at the hydrogen. This will cause the hydrogen to undergo a rapid increase in temperature until it reaches the point where it *ionizes* and becomes plasma.[5] How hot must the hydrogen become before this happens? It will not need to reach anything like the 200 million degrees Fahrenheit required for thermonuclear fusion. Nor need it reach even 40,000°, the temperature at which *all* substances ionize and become plasma. A temperature of only about 10,000° is necessary to ionize hydrogen.

A large fleet of spaceships acting as magnetic scoops will have to move through the Orion Arm, pushing iron, nickel, hydrogen plasma and other ionized interstellar material before them. And to achieve this ionization, each scoop-vehicle will bombard the interstellar medium with the most powerful laser beams that the scientists of that distant age will have been able to devise.

Using magnetic fields in space as giant interstellar bulldozers to accumulate matter will be a similar task, but one in essence much simpler than achieving a fusion reaction inside a closed container. It will not matter in

*A literal translation: 'Nature is only conquered by obedience.'

the least if a small part of the interstellar material escapes, so long as the greater part is gathered up. The idea of a magnetic scoop in space has been thought of before, but for a different purpose. The concept will be familiar to many people who have considered the feasibility of interstellar travel; I do not mean those few who have discussed instantaneous journeys, but the many who have long brooded on the problem of how to accelerate a spacecraft to a speed close to that of light. The engineer Robert Bussard pointed out in 1960 that even with the most efficient means of propulsion yet conceived of, no spacecraft would be able to achieve such a speed because it would need to carry millions of times its own weight in fuel. Bussard worked out an ingenious solution to this problem, which has become known as the 'Bussard Ramjet'.[6]

This ramjet engine would operate on a similar principle to the air-breathing jet aircraft engine, which sucks in air in the front and expels it in the rear. Bussard's ramjet, to be used on an interstellar spacecraft, would carry in front of it huge magnetic fields, thousands of miles in radius, which would scoop up particles of ionized hydrogen as it passed through space. The collected hydrogen would enter the fuel-tanks and power the spacecraft through a fusion engine. Many writers have since developed different versions of this concept.[7] Significantly, none of them have forgotten in their engineering enthusiasm that such a ship would be carrying people, cargo, life-support systems and delicate communication instruments, and that there would need therefore to be strict limits on the quantity of hydrogen collected each second. According to one view, the intake of hydrogen into the fuel-tank ought not to exceed 0·0001 ounces per second. Whatever the safety limit may be, the dangers of exceeding it are obvious. The fuel-tank would become overloaded. Temperatures in all parts of the spacecraft would begin to rise. If the ship was moving at close to the speed of light, it could be considered as being stationary while dust-grains and hydrogen atoms crashed into it at close to the speed of light. Radiation would penetrate the ship with an intensity 100,000 times greater than the intensity of sunlight at the Earth's surface. It is easy to predict what would happen: the journey would never be completed because the ship would fragment from overheating, and the people in it would be fried.[8]

But this difficulty only arises if the ship contains people and all the paraphernalia, in the form of special instruments, living quarters, and air and food production which people require. It will be possible to construct a black hole of ten Solar masses at a distance of about one light-year from the Sun, using the same basic method as that of Bussard's ramjet.*

Now the ramjet system, as envisaged by Bussard, is inherently inefficient in the sense that it would fail to achieve its aim, namely to accelerate a ship to a speed close to that of light. The reason is obvious: according to the engineers who have studied the problem, only about 1 per cent of the interstellar material can be used as fuel. The rest of it piles up in front of the vehicle. Acceleration becomes increasingly more difficult, for the ramjet engine is expending energy pushing its way through the plasma (which the ramjet has itself accumulated) which it ought to be expending on accelerating the ship. It is unlikely therefore that a simple ramjet engine would ever succeed in accelerating a ship to anything more than a low percentage of the speed of light.[9]

But a low percentage of the speed of light will be perfectly suitable for the purpose of accumulating matter and bulldozing it forwards. Ships will be launched into interstellar space with no other purpose but to collect at different vantage points the material for constructing the black hole. Nobody will travel on these ships, not

*This size, of ten Solar masses, is probably ideal. If it was much smaller, the navigable aperture would be too narrow to admit spaceships; and if it was very much larger there would be a definite risk to the stability of the Solar System. A simple equation for calculating the width of the navigable aperture from the original mass of the black hole, with an accompanying table is, given in Appendix I.

only because the expense of maintaining people on board would multiply the cost of the project by a factor of many thousands, but because anyone on board would certainly be sucked with the ship into the black hole and crushed to death when the actual formation of the hole began. These ships will be suicide vehicles, with no other purpose than to accumulate matter. Everything on board will be subordinated to that purpose; they will carry no scientific instruments except those necessary for their mission. The operation of their engines will be fully automatic. Their communication antennae will be as simple and as compact as possible, the only necessity being that their courses can be changed by distant remote control in an emergency. There need be no restriction on their rate of accumulation of matter. To restrict this rate of accumulation to a rate of 0·0001 ounces per second would be pointless, since at that rate the process of constructing the black hole would last longer than the future age of the Universe! Instead, their task will be to accumulate matter at the fastest possible rate.

Some further practical questions must now be discussed: what kind of ships will be needed? Where would they be constructed? And above all, how much interstellar material could they accumulate and drive before them in how short a time?

Long before the twenty-third century, when mankind may be wealthy enough to carry out such a project, many other less grandiose engineering schemes in space are likely to have been put into effect. A good example is the proposal of Professor Gerard O'Neill of Princeton that cylindrical cities can be constructed in space, perhaps early in the twenty-first century, which will provide homes for tens of thousands of people.[10] Laymen often make the ill-informed objection: 'These schemes are absurd, because the cost of launching into space the materials for this construction would, in any age, be quite prohibitive.' Indeed, nobody in his senses would dream of trying to launch into space, from the Earth's surface, materials weighing millions of tons. There will never be any need to do so. Almost limitless amounts of suitable material was 'launched' more than 4,000 million years ago, when the Solar System was formed. Between the orbits of Mars and Jupiter is a great swathe of tiny planets, known as the Asteroid Belt. Numbering about 50,000, these unevenly shaped lumps of rock range from Ceres, the largest, which is about 430 miles across, to much smaller fragments a few hundred yards in diameter.

Consider a typical asteroid, Eros. It is a tiny planet about fifteen miles across. The only reason that its dimensions are at all well known is the eccentricity of its orbit. In 1931 and in 1975, it approached to within 14 million miles of the Earth, making it temporarily the nearest astronomical body to us beyond the Moon. But apart from this orbital eccentricity, in terms of size Eros is a typical asteroid. There are thousands of others of roughly the same diameter to be found in the main body of the Asteroid Belt. Some of these medium-sized asteroids can be adapted to serve as magnetic ramjets. Asteroids have often been proposed as raw materials for space engineering projects. Being of extremely low mass compared with planets, they will be much easier to explore and to redesign. This is because the surface gravity of an asteroid is far weaker even than the one-sixth gravity of the Moon. A man who weighed 150 pounds on Earth would weigh a mere 4 pounds on Eros, and not more than 1 ounce on some of the smallest asteroids. The very shapes of many of the asteroids make them suitable for high technological purposes. Instead of being spherical, like stars and planets, they are oblong and roughly cylindrical.[11]

It may prove a much easier engineering feat to hollow out an object of cylindrical shape, and make it into a ship, however jagged and uneven it may be, than it would be to do the same thing with a sphere. If Eros, which has been often observed during its close visits to Earth, is of cylindrical shape, it may be regarded as certain that many of the other thousands of asteroids which do not have such wandering orbits, and which cannot yet be closely observed, will be of the same general form. They will therefore be perfect raw material for the construction of magnetic bulldozers. (The tiny moons of Mars bear this out, for they are jagged, and non-spherical.) A mass of literature exists on the feasibility of mining and redesigning asteroids, all of which suggests that being merely great lumps of iron and nickel, and being of no economic use to mankind in their present orbits, the asteroids may be adapted for many profitable uses.[12]

We can now envision perhaps a thousand gigantic unmanned ships, being redesigned asteroids, accelerating out of the Solar System in slightly different directions as they follow a rigid mathematical plan. They will be travelling, by different routes, to a point one light-year from the Sun, on a direct line with the central regions of the Galaxy, where the black hole is to be constructed.* It will be a long project. I have written somewhat glibly of the 'ten Solar masses' of interstellar matter that must be collected; yet few people will comprehend what a gigantic amount of material is equivalent even to a single Solar mass. It may convey little simply to call the mass of the Sun 2×10^{27} tons (2 followed by 27 zeros). It might sound more vivid to say instead that the Sun is 330,000 times more massive than the Earth. Ten Solar masses (2×10^{28} tons) is ten times this amount, and so the interstellar bulldozers will have to collect material whose total mass is more than 3 million times the mass of the Earth!

The task of the ships, guided by an intricate computer programme which directs each one of them, will not be quite so difficult as this huge figure might imply. The magnetic fields of the ships will start pushing matter

*This distance of one light-year, 6 trillion miles, a distance nearly 1,700 times greater than that which separates us from Pluto, our furthest planet, may seem a long way for the unmanned ships and for human voyagers to follow after them when the project has been completed. But to build the black hole nearer to the Sun than one light-year might be impossible, or at least much more difficult, as I explained at the end of the last chapter, because there would be insufficient quantities of interstellar material.

before them as soon as they are beyond the orbits of the outer planets and thus clear of the Solar System. Now the basic inefficiency of the Bussard ramjet, it will be remembered, is caused by the fact that only about 1 per cent of the hydrogen plasma which it collects gets into the engine which provides its thrust. The rest of the hydrogen, which the magnetic field has attracted piles up in a huge cloud stretching tens of thousands of miles in front of the vehicle, inhibiting its acceleration. But this 'piling up' of interstellar matter is exactly what the black hole bulldozers will be trying to achieve!

Let us assume that the magnetic field generated by each vehicle will be about 150,000 miles wide, and that the electrical energy to generate the field can be supplied through a superconducting system powered by the vehicle's main engine. The iron, the nickel and the hydrogen plasma, piling up in front of the magnetic field, will form a great column of matter stretching forwards for hundreds of thousands of miles. This mass will be propelled forwards by the movement of the vehicle. In these conditions, it will form a great eddy, or current, generated by its own swirling movements around the magnetic field, similar, perhaps, to the eddy effects of wind in a narrow rock chasm.

Figure 6 indicates, in the simplest possible way, how such a vehicle would work (it is not to scale).

We can calculate that the interstellar matter will be 'herded', if such a word is permissible in the context, into a new direction at a rate of thirty-five ounces per second per ship; or, if a fleet of 1,000 ships is at work, at a rate of 35,000 ounces: about one ton per second. Suppose, that, because of the great distances in a sphere round the Sun with a radius of one light-year, twenty years elapse between the time that the ships embark on their mission and the time when they begin to accumulate matter in significant quantities. After this period, however, another factor will be working in their favour; the rate of increase in the accumulation of matter will itself increase.

The ionized matter will itself ionize fresh matter,

which will in turn be caught up in the eddying forward movement. Imagine a cowboy herding cattle in what the cigarette advertisements call 'Marlboro country'. The herded cattle persuade other cattle, by their example to run in the same direction as themselves, and their herd instincts thus ease the cowboy's task. So it will be, in a matter of speaking, with the ionizing process and the accumulation of interstellar matter. It is not unreasonable

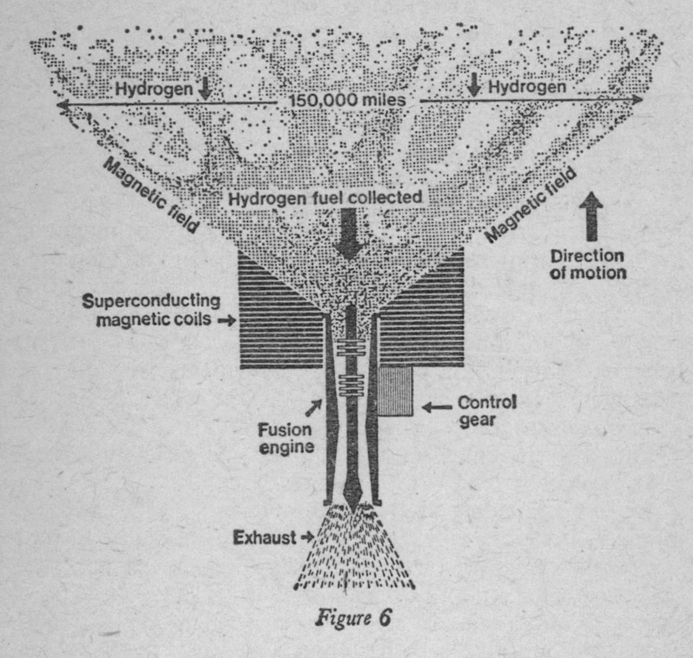

Figure 6

to predict that when this process of accumulation really begins to mount, that its actual rate of increase will itself increase at a rate of perhaps 1 per cent every twenty-four hours. Now anything that increases at 1 per cent per day will double every seventy days. If such a rate of doubling can be achieved, with the ships cruising at about 7 per cent of the speed of light, or 47 million miles per hour, a black hole can be constructed about fifteen years after the end of the twenty-year preparation period, making a total of under forty years between the launching of the ships and the formation of a black hole.

Yet some people may object that the formation of a star, like the Sun, takes tens of millions of years, even after the materials which are going to comprise that star have assembled in one region of space; and so why would black holes be formed so much more quickly? The answer is that black hole formation and star formation are quite different processes. A black hole, being the immediate consequence of the accretion of a minimum of three Solar masses, will assume its final shape in *less than a second*. As Kip S. Thorne of Caltech explains this, 'when a black hole first forms, its horizon may have a grotesque shape and may be wildly vibrating. Within a fraction of a second, however, the horizon should settle down to a unique smooth shape.'[13] And so it is not unrealistic to allow only forty years for the completion of the project.

Forty years, however, is a very long time in the history of a human society. The political problems inherent in the construction of a black hole may prove more difficult to solve than the engineering ones. Even two and a half centuries hence, when mankind is likely to be more than 1,000 times wealthier in real terms than it is now, only governments will be rich enough to finance the construction of a black hole.* What arguments can the 'cosmic engineers' put forward to persuade a government—or an association of governments—to finance the project? And even assuming that they obtain enough funds to begin work, will they be able to obtain more with which to continue? Absolute candour on their part at too early a

*This will not always be true. Six centuries hence, for instance, assuming a reasonably steady economic growth rate. any reasonably well-off private corporation will be able to afford such an expenditure. But in the poorer days of the twenty-third century, only governments will have the money.

stage could have disastrous results; they will have to acquire the technique of concealing vital information without appearing to do so. And they will need, above all, to be skilled psychologists.

7
The Politicians

Forty years! It is a long time for the duration of a relatively expensive, government-funded project which does not give mankind the slightest advantage during that period. It would be quite out of human character, if during that time, the project did not attract the hostile attention of a number of politicians.

It will make no difference whether the political system of the day is democratic or authoritarian. The rules of politics are not changed by constitutions, however much their outward forms may differ, and an ambitious man can always hope to win votes or favour by 'exposing' an allegedly scandalous situation. It will sound plausible to oppose the black hole project on three separate grounds: that it might be dangerous to mankind; that the money should be spent on more urgent terrestrial or interplanetary projects; and that its managers are suspected of having consistently lied to the government. A black hole of ten Solar masses at a distance of one light-year from the Sun could not in reality pose the slightest threat to mankind, or to any part of the Solar System, because it is too small and too far away; but this fact, which any mathematics student could verify in minutes, is likely to be swept away by emotion. We know from the campaigns against supersonic passenger aircraft in recent years that influential people are sometimes undeterred by simple scientific facts. In this case, these people will imagine that they have an easy prey. The very phrase, 'black hole' has an unpleasant and menacing ring. Those scientists who dislike the project will be able to remind an eager public that black holes are 'cannibal stars' that 'devour' all matter in their vicinity. It will be a poor politician who cannot make capital against the project at election times by using the word 'devour' to conjure up the image of the primeval forest.

Yet the cosmic engineers will have some advantages in this political battle. There is, perhaps happily, an enormous difference between a humanist and a scientific education. Since 1868, when T. H. Huxley reported that one could graduate with the highest honours from Oxford or Cambridge without ever hearing that the Earth went round the Sun, this difference has widened.[1] C. P. Snow told an interesting anecdote in 1959. A. L. Smith, a distinguished professor of classical history, a man much given to convivial gossip, was invited as guest of honour to dinner one night at the high table at Trinity College, Cambridge. Snow told the story like this:

> He addressed some cheerful chit-chat to the man opposite to him, and got a grunt. He then tried the man on his right and got another grunt. Then, rather to his surprise, one looked at the other and said, 'Do you know what he's talking about?' 'I haven't the least idea.' At this, even Smith was getting out of his depth. But the President, acting as a social emollient, put him at his ease by saying 'Oh, those are the mathematicians! We never talk to *them*.'[2]

A. L. Smith would probably have had much better conversation from an astronomer, or a physicist or an engineer since pure mathematicians do have a tendency to be exclusive. But Snow nevertheless had a powerful argument for the existence of two separate and widening cultures, the humanist and the scientific. He very properly deplored this separation, pointing out its dangers for the survival of industrial civilization: how could the humanist leaders of these civilizations, presumably ignorant of the

Earth's orbit round the Sun, possibly be suitable managers of an industrial technology that depended upon science?

Yet Snow perhaps takes a too gloomy view. It may be that the ignorance of the humanist politician about science is greater than the scientist's ignorance about politics. I have met many politicians who had not the faintest idea of the difference between a planet and a star. On the other hand, I have seldom known a scientist who did not have some idea of the character of the political parties in his country, and for which he was going to vote. If this represents a general truth, then the existence of Snow's 'two cultures' is not something which the promoter of a high technological project ought to deplore; it is something which he may on some occasions be able to exploit.

There is no point in softening words. If the managers of the black hole project want to win their political struggle to obtain and keep the funds necessary for their project, it will be idle for them to imagine that they could ever be victorious in an open, wholly honest public debate. They would never be able to withstand the waves of ignorant, populist emotion. They cannot win the argument by open means; they will have to win it by subterfuge.

Imagine, therefore, that by the end of the twenty-second century, the managers of the project have succeeded in obtaining public funds for what is officially termed a 'pilot controlled experiment to determine whether the construction of a black hole is feasible'. After several years have elapsed, the politicians perceive dimly that the project managers are not restricting themselves to a mere 'pilot experiment'; but instead they are carrying out the main project. This discovery is followed by popular indignation, fuelled by the suspicion that the scientists are acting without authority and perhaps even endangering the survival of mankind. At length, after prolonged clamour, the people of the early twenty-third century may one day read the following news item:

Environment Commissioner Bandwagon, noting what he called 'extreme public concern', today ordered the Black Hole project to be suspended pending the outcome of a full public enquiry into the project, to be chaired by himself.

Our political correspondent believes that the Commissioner's decision is not unconnected with his plans to seek higher office in the coming elections. Campaigning with the populist slogan, 'Dynamic Social Progress', he is determined to present an image of . . .

Our science editor says that whatever the findings of the inquiry, the Black Hole project will be set back for at least a generation, since the magnetic scoops will have lost all impetus and direction. Each day that the dust-clouds lose ionization, it will be progressively more difficult to . . .

We will move without delay into the first and decisive day of Commissioner Bandwagon's public hearings. With an excited crowd of his supporters in the public gallery, all sharing his determination to suppress the Black Hole project, he confronts the principal witness, the project's chief administrator, whom I will call Dr Black.* Their dialogue may go something like this:

COMMISSIONER BANDWAGON: Good day, gentlemen, and hello to you folks in the gallery. Dr Black, you will be aware of the Departmental order which shuts down the Black Hole project pending the outcome of this inquiry. Just as a matter of routine, can we have your assurance that your organization has complied with the order, and that the magnetic fields have been switched off?

DR BLACK: Gentlemen, let me say one thing first of all. My people are loyal civil servants. We will always do our utmost to comply with orders from your depart-

*I make no apology for using fictitious characters to illustrate an argument. Plato did it, as did Galileo. (It was Galileo's obstinate character Simplicio, in his *Dialogue* of 1630, who made Pope Urban VIII imagine that he himself was being parodied. The infuriated Pope at once became the astronomer's bitter enemy. Patrick Moore's *Watchers of the Stars* (Michael Joseph, London, 1974) gives a fascinating account of the ensuing confrontation.)

ment or from any other. But I regret to have to inform you that in this case there are special circumstances which prevent us from complying.

COMMISSIONER BANDWAGON: I cannot have understood you correctly. Are you seriously telling this Committee that you intend to disobey a direct Departmental order?

DR BLACK: Our intentions have nothing to do with it. Our policy is to obey scrupulously all orders which we receive from any branch of the government. But we cannot perform something which is physically impossible.

COMMISSIONER BANDWAGON: What do you mean, physically impossible? Why is it impossible to send the appropriate radio messages which will switch off the magnetic field?

DR BLACK: Because, Commissioner, we are no longer dealing with a magnetic field. We are now dealing with a gravitational field. I agree with you that a magnetic field can be switched on and off like an electric light. There would be a certain time delay before our messages reached the giant magnetic scoops, because of the limiting speed of light, but that is another matter. But a gravitational field is a wholly different question. Nobody, neither you nor I nor the President himself, can switch off a gravitational field at whim. The last politician who tried to switch off a gravitational field was King Canute of England, and he only succeeded in getting his feet wet.

COMMISSIONER BANDWAGON: Come now, Dr Black. I would hate to feel that you were deliberately wasting the time of the Committee for some devious political purpose. We have always been told that your black hole is being constructed by means of a magnetic field. Why is it not possible simply to switch off that magnetic field?

DR BLACK: Until about nine months ago, sir, the mass of our potential black hole was being accumulated by magnetism. But at about that time, the mass became so great that matter outside it began to be accreted by gravitation. Our magnetic fields, which started the process of accumulation of matter, then became insignificant. To put it simply, about nine months ago, because our operations involved gravitational collapse, the magnetic force became completely overwhelmed by gravitational force.

COMMISSIONER BANDWAGON: What! Why was my Department not informed of this fundamental change in your operational procedure?

DR BLACK: With respect, Commissioner, that is a very naïve question from a politician. Nine months ago, if you had ordered the project to be shut down, we would have had to comply, since a magnetic field can easily shut down. But my people have perhaps considered the long-term prospects of the human race more profoundly than anyone in this committee-room, and we did not wish to shut it down. We believe that the construction of a black hole will be a priceless boon to humanity, and so we thought it advisable not to publicize this information in case it might be misused when election time came round.

COMMISSIONER BANDWAGON: And so you lied?

DR BLACK: Choosing not to make unsolicited statements is hardly the same thing as lying.

COMMISSIONER BANDWAGON: I still say you have lied! I have here in front of me a copy of your original charter. It states that your agency is authorized to spend an annual sum on what is termed a 'pilot controlled experiment to determine whether the construction of a black hole is feasible'. In no way does that wording permit you to go ahead and construct a black hole. You have far exceeded your authority. And by your own admission here today, you have deliberately withheld information from my Department to which it was entitled. I shall inform the Public Prosecutor that your conduct amounts to nothing less than a criminal conspiracy. (*Applause from the gallery.*)

DR BLACK: Tell the Public Prosecutor whatever you please. My conscience is clear. We were authorized to carry out a 'pilot-controlled experiment'. My predecessor who accepted those instructions was an electrical engineer, as I am. To an electrical engineer, the words 'pilot-controller' mean the same thing as 'master-controller', a controlling mechanism which overrides all subsidiary systems. Perhaps the word 'pilot' misled you. Perhaps you were thinking, like domestic householders used to, of the 'pilot flame' in their old gas cookers which was subsidiary to the main cooking flame. But in an electrical system, or for that matter any kind of organizational system, the phrase, 'pilot control' has a wholly different meaning. Let me give you an obvious example. On a spacecraft or an aircraft, there is usually someone on board called the 'pilot'. He is no apprentice or junior adviser; he is the captain, the chief officer of the vehicle. No, sir; an electrical engineer can act only upon the technical and literal meanings of words, and not upon their popular meanings. Your error was to imagine that we were working on a 'pilot project', which I admit would have meant quite a different thing. You mistook scientific language, which is precise, for technocratic jargon, which is not. Allow me to buy you a scientific dictionary for Christmas.[3]

COMMISSIONER BANDWAGON (almost shouting): But you said it was only going to be an experiment—

DR BLACK: And so it was. And I am proud to announce that the experiment shows every sign of producing a positive result. The proof of that is that a black hole is being successfully constructed.

COMMISSIONER BANDWAGON: (after a pause): This is outrageous. You are setting yourself up above the Government. Who gave you the right to decide what is good or bad for humanity? Are you not aware of the great number of distinguished scientists who have attested that the formation of a black hole so near to the Sun might endanger all life on Earth and in our colonies on the inner planets?

DR BLACK: Frankly, sir, the number of truly distinguished scientists may be numbered on one hand, and not one of these, to my knowledge, has voiced such eccentric opinions.

ANOTHER COMMITTEE MEMBER: You know, Dr. Black, the way you talk really annoys me. I understand that this black hole of yours is going to be ten Solar masses—ten times more massive than the Sun. You cosmic engineers talk about 'ten Solar masses' in the same glib, casual way that you might speak of ten tubes of toothpaste or ten packets of cigarettes. Have you really considered the size of this monstrous, evil thing that you are building, so near to the Sun? It will be *three million times* more massive than the Earth! How can you possibly sit there and pretend that you have calculated that it will not wreak any destruction on us? Dr Black, do you want to go down in history as the man who wiped out the entire human race? (*Frenzied applause from the gallery.*)

DR BLACK: Nobody is going to be wiped out. Please allow me, gentlemen, to explain the current situation in terms so simple that even a politician can understand it. Consider the floor of a room. There are only two conditions which that floor can be in; it can either be clean or dirty. Setting aside the effects of earthquakes or explosions, there is no third condition. Either there is dirt or there is no dirt. It is the same with interstellar space, which must either be dirty or clean. A thousand years ago, or maybe 10,000 years ago, nobody really knows, the Solar System entered a dirty region of space which we call the Orion Arm of the Galaxy. Our policy is to scoop up a portion of that dust-arm so that it accretes at a safe distance of about one light-year from the Sun, in order to create a black hole of ten Solar masses. Now you have thrown a big statistic at me, so I am going to throw one back at you. The black hole is going to be situated a whole light-year away. Do you know how far that is? Let me just give you an idea. When you fly to the Moon on your political junkets, your maximum cruising speed is 25.000 miles per hour. You probably think that's pretty fast, but by astronautical standards it's a snail's pace. Do you know how long it would take you to travel one light-year going at 25,000 miles per hour? It would take you no less than 27,000 years! That's what I mean by a safe distance from the Sun.

A lot of second-rate scientists, of the kind who seem to live only to pontificate on television, have been making their usual wild statements about what might happen to those parts of the Orion Arm *that lie between us and the potential black hole*. These people have the idea that the magnetic field might somehow get out of control, and that the Earth and the Sun might be partly sucked into the black hole, putting an end to human life. But nothing like this could possibly happen. The reason is very simple; the accretion of matter to a single dense point must leave all around it a great swathe of empty, clean space whose density is being reduced to that of normal interstellar space outside the Orion Arm, where there is only one atom of matter per cubic centimetre. So the Earth cannot be wrenched from its orbit, or be made to change its shape, because there is no longer any matter in the near-by interstellar medium to exert any unusual magnetic or gravitational field upon the Earth.

I will prove this paradox with another. If this were not the case, and if the black hole *was* going to exert a force upon the Solar System, *the effects of this force would already be apparent*. But no such effects have been observed. Our knives and forks have not hit the ceiling. There has been no increase in the usual frequency of earthquakes. The Magnetic North Pole has not shifted to any marked degree. The planets are orbiting the Sun at their usual distances, and the Sun itself is not displaying any peculiarities.

Please remember, gentlemen, that we have been planning this operation for more than sixty years, and that people have been thinking about it for more than a century. We *knew* that there would be no dangerous side-effects, because we *calculated* that there would be none. We lied to no one. We kept nothing secret. We communicated all our ideas through the technical journals, which you politicians apparently did not bother to read. You complain because we did not give monthly news conferences, but why should we? We are scientists and engineers, not public nursemaids for lazy minds. The real scandal in this affair is not any lack of candour on our part, but the quite unnecessary alarm and hysteria which you are spreading through the world by these ridiculous committee-hearings.

COMMISSIONER BANDWAGON: I can hardly find words

to answer you. You sneer at us for being mere politicians, because we cannot read journals which are filled with technical jargon and advanced mathematics that no humanist can understand. You act upon the most obscure meanings of simple words, you evade the will of my Department by chicanery, and without consulting anyone, you arrogate to yourselves the right to decide our future. *Our* future, which you have robbed us of all power to shape! Aren't you even worried by the appalling impression you are making today on public opinion?

DR BLACK: On current public opinion? Not in the least. Historians will take a more charitable view, especially since they will be writing from the vantage point of a Galactic community, opened up to them by our projects. Nine months ago, when the magnetic fields were still bulldozing matter, I was very anxious lest public opinion should become misinformed about what we were doing. But now I snap my fingers at it. I can afford to. We can now sit back and let nature finish the job for us. Don't look so disheartened, Commissioner! We are giving humanity the chance to colonize millions of habitable but uninhabited worlds. Doesn't that sound to you like 'dynamic social progress'?

But Dr Black, having concealed from everyone his knowledge of the exact moment when the quantity of accumulated mass would reach a significant percentage of a Solar mass, whose gravitational field would swamp the magnetic fields, is still not quite happy with all aspects of the project, even after his triumph over the politicians. For there is one section of his own organization whose work he only dimly understands. The people in this section are developing a project which would infuriate Commissioner Bandwagon even more—if only he knew about it. They are trying to solve the problem which must have occurred to many readers of this book: even if astronauts succeeded in passing through the black hole and thereby making an instantaneous journey to another part of the Galaxy, *how will they be able to return?* In view of cosmological terminology, there would be only one possible name for the administrator of this parallel project. If I were going to write about him (which I shall do as little as possible) I would have to call him Dr White.

8
The Other End of the Labyrinth

The national flag of South Korea displays in its centre a circle consisting of two curving halves that are coloured respectively red and blue. The two halves invade each others' areas in such a way that these areas of the circle are equal in size. This pattern is an eastern symbol of unknown antiquity. The blue and red areas are called respectively Yin and Yang. These symbols represent all the dualities of nature—good and evil, pain and pleasure, light and darkness, male and female, hot and cold, positive and negative—and so forth.[1] This ancient symbol illustrates a principle over which Oriental philosophers have brooded for thousands of years, the apparently flawless *symmetry* of nature.

Ever since the publication of Isaac Newton's Third Law (mainly applicable today to the operation of rockets and jet engines), that 'every action has its equal and opposite reaction', the principle of symmetry has had an equal fascination for the scientists and the creative artists of the West. In architecture, this fascination goes back long before Newton at least to the building of the Parthenon. From that time onwards, it has been found that a symmetrical structure whose length is 1·62 times its height produces for some strange reason an effect that is especially pleasing to the human eye.*[2] In fiction and in poetry, the effects of symmetry are equally admired. A good example is Edgar Allen Poe's story, *The Purloined Letter,* where the answer to the mystery is that there never was a mystery at all. The blackmailing letter for which the police had ransacked an apartment, taking every item of furniture to pieces, was all the time lying openly on a desk, where no one ever thought of looking for it. We get the same dramatic effect from those epigrams or poems that are 'palindromic', that read backwards just as they read forwards. One example is a verse in which the lines read similarly in either direction:

> As I was passing near the jail
> I met a man, but hurried by.
> His face was ghastly, grimly pale.
> He had a gun. I wondered why
> He had. A gun? I wondered . . . why,
> His face was *ghastly*! Grimly pale,
> I met a man, but hurried by,
> As I was passing near the jail.

Or again, when the very letters are also transposed, we have the anonymous epitaph to Ferdinand de Lesseps, builder of the Panama Canal:

A man, a plan, a canal, Panama!†

*This architectural form is known as the Golden Rectangle. The number 1•62 or, to be more exact, 1•61803, is obtained by halving the sum of 1 and the square root of 5. It is also the ratio between any number above 5 in the so-called Fibonacci series of numbers, in which each number is the sum of the previous two, for example, 1, 1, 2, 3, 5, 8, 13, 21, 34, 55, 89, 144, etc. If we divide any Fibonacci number by its predecessor, i.e. 55 by 34 or 144 by 89, we obtain an answer which is progressively nearer to the Golden Mean 1•61803 as the numbers divided become higher.

†There are many other such palindromes. Adam might have introduced himself to Eve with the words, 'Madam, I'm Adam.' Or again, the remark attributed to Napoleon: 'Able was I ere I saw Elba.' A good words-only palindrome is: 'You can cage a swallow, can't you, but you can't swallow a cage, can you?'

These agreeable symmetries are matched in nature, where physicists have established the existence of 'anti-matter' as an opposite to matter. Their discovery dates from 1932, when Carl Anderson saw in a cloud-chamber at Caltech the path of an electron which was moving in the opposite direction to that in which electrons ordinarily moved. He called it a positron, or anti-electron. It has been found since then that all particles have their anti-particles, of opposite electric charge, and that when two opposites collide they both annihilate each other in a tiny explosion. Elsewhere in the Universe, it has therefore been reasoned, there must exist 'anti-galaxies', composed entirely of 'anti-matter', in which every atom is an 'anti-atom' to the corresponding atom in our own Galaxy. I do not mean, of course, that an anti-galaxy would be a precise mirror image of our Galaxy, containing a parallel Earth and a parallel Europe and America; that would not be science, but magic. I mean only that the other galaxy would be anti-matter, and that if ever the two galaxies, or any matter from either of them, came into collision, both would be destroyed in a violent explosion in which matter was converted into energy in its entirety, in accordance with Einstein's famous equation, $E = mc^2$.* Dr Edward Teller gave a lecture in 1956 in which he discussed the great energies which contact between matter and anti-matter would produce. Another poem gave the amusing reply which the *New Yorker* made to Dr Teller a few weeks later:

> Well up beyond the tropostrata
> There is a region stark and stellar
> Where, on a streak of anti-matter,
> Lived Dr Edward Anti-Teller.

*The equation means simply that the energy of the explosion, E, usually expressed in ergs, equals all the mass involved in grams, multiplied by the square of the speed of light in centimetres per second. In other words, a very small amount of mass, if converted at 100 per cent efficiency, can produce a gigantic amount of energy. Conversions of much lower efficiency provide the reactions of nuclear bombs today.

Remote from Fusion's origin,
He lived unguessed and unawares
With all his anti-kith and kin,
And kept macassars on his chairs.

One morning, idling by the sea,
He spied a tin of monstrous girth
That bore three letters: A.E.C.*
Out stepped a visitor from Earth.

Then, shouting gladly o'er the sands,
Met two who in their alien ways
Were like lentils. Their right hands
Clasped, and the rest was gamma rays.[3]

An anonymous limerick, although written in the nineteenth century, long before the discovery of anti-matter makes the same point even more succintly:

There once were two cats of Kilkenny,
Who thought there was one cat too many,
So they mewed and they bit
And they scratched and they fit,
'Til, excepting their nails and the tips of their tails,
Instead of two cats there weren't any.[4]

The principle of symmetry can be extended further. Nearly every natural phenomenon has its 'anti-phenomenon', in which events are reversed either in space or time; the tides advance and recede, there are explosions and implosions, and people (and perhaps some animals as well) are left- and right-handed. Planets appear to travel anti-clockwise round the Sun, but that is only true for an observer looking 'down' on the Solar System from 'above'. Another observer looking 'up' at it from 'below' would see them moving clockwise. Who is to say which is the 'correct' and which the 'incorrect' way to look at the Solar System? A spiral or an elliptical galaxy, as one can see from the pictures in the Hubble

*Atomic Energy Commission. This was before the days of NASA.

Atlas mentioned earlier, is symmetrical; if all the pictures of such normal galaxies in the Atlas were printed back to front, as in a mirror, few people would be able to detect the fraud. Unless it is irregular or unstable, a galaxy, like a star or a planet, is superimposable on its mirror image. It is even conceivable that the entire Universe is superimposable on its mirror image. One physicist, P. C. W. Davies, has proposed that in another inhabited Universe, if such an idea can be imagined, time might run backwards instead of forwards.[5] But the people in this universe would see nothing abnormal in this state of affairs. To them, a universe where time ran *forwards* would seem as bizarre and as inexplicable as theirs might seem to us. There could be no basis for an agreement between the peoples of the two universes that 'forwards' mean tomorrow, and that 'backwards' means yesterday. The very terms, 'past and future', 'left and right', 'north and south' are arbitrary in the sense that both peoples will have defined them at random.

There is an interesting problem of modern physics on these lines known as the Ozma Problem, after the fabulous Land of Oz. It goes like this: Supposing you had established communication by radio with a being on another planet, and built up a common language with him, how would you make him understand what you meant by 'north' and 'left'? Until the late 1950s, the problem appeared insoluble. It seemed that it would be futile to ask the being to perform an experiment to synchronize his north with your north, and his left with your left because, so it seemed, there could be no such experiment. The north and south poles of his planet, and the left and right sides of his body, would be named as arbitrarily as yours. It would be useless to say to him: 'My heart is on my left side', because he would have no idea what you meant by 'left'. The Ozma Problem was in fact solved in 1956 (and one physicist, Richard Feynman, danced a jig when he heard the news) through the discovery that the atomic nucleus of the radioactive isotope cobalt-60, which emits a continuous stream of electrons, emits more electrons from its south pole than it does from its north.[6] It would therefore be possible for the alien being to perform this experiment on your instructions and learn from it your north and left. But apart from cobalt-60, no other known substance is sufficiently unsymmetrical to solve the Ozman Problem.

But surely, it will be said, the rotating black holes which litter the Galaxy are themselves unsymmetrical. Are they not phenomena for which there are no anti-phenomena? For black holes are essentially unsymmetrical. They are stations for one-way traffic; matter flows into them but cannot flow back. They implode but cannot explode. Yet this view of the cosmic situation is itself unsymmetrical and must therefore be incomplete. In our Galaxy alone, it has been estimated that black holes devour matter at an approximate rate of one Solar mass per day, and that consequently they must have been devouring matter at this rate for billions of years in the past.[7] Now the age of the Galaxy is generally accepted as being about 10,000 million years. If, therefore, black holes had started devouring matter at this pace soon after the Galaxy's formation, the Galaxy would have ceased to exist after about a mere 270 million years, every gram of its matter having been devoured.

This situation is plainly absurd. The Galaxy is still filled with thousands of millions of shining stars. There has to be a compensating mechanism which pours matter back into the Galaxy as fast as the black holes suck it out. The great astronomer Sir James Jeans reached this conclusion, or something like it, as long ago as 1928. Writing about other spiral galaxies in the Universe, he said:

> Each failure to explain their spiral arms makes it more and more difficult to resist a suspicion that the spiral nebulae are the seat of forces entirely unknown to us, forces which may possibly express novel and unsuspected metric properties of space. The type of conjecture which presents itself, somewhat insistently, is that the centres of the nebulae are

of the nature of 'singular points' at which matter is poured into our Universe from some other spatial dimension.[8]

Seven years later, as we have seen, a version of this idea, the cosmological basis for the modern theory of black and white holes, was presented in rigorous mathematical form by Einstein and Rosen. It is now widely considered that for every black hole created there must come into existence, instantaneously, a corresponding *white hole*.

A white hole is no more strange an object than a black hole. It is simply its opposite. A black hole is an implosion, and a white hole is an explosion. Nothing can escape from a black hole; *everything* must sooner or later escape from a white hole. If black holes can be thought of as phenomena, then white holes are their corresponding anti-phenomena. The theorist Robert Hjellming made this point in a famous article in 1971 when he declared: 'We can say that black holes are related to white holes because, at certain points, our Universe is multiply-connected through black hole–white hole singularities. In other words, the black holes of our Universe could supply our white holes.'[9] Sir Fred Hoyle puts this idea just as plainly: 'The structure developed by a local collapsing object consists, not just of a black hole, but of a black hole together with a white hole.'[10]

Now it has been very difficult for observers to discover and pinpoint black holes, for the very good reason that neither light nor anything else escapes from them. Today, this is only possible when the black hole is one of a pair of stars, in orbit round each other, and the second and visible star is seen to 'wobble' in its orbit, or to emit violent outbursts of X-rays.* But white holes must be

*The next generation of optical telescopes, which are likely to be used in space, away from the Earth's befogging atmosphere, may be able to see something even stranger than this: it may be possible to see through them the shape of the black hole's companion star being warped into a pear-shape by the ferocious gravitational tug of the hole. But such observations are not likely to be possible until the late 1980s, when optical telescopes are used in space.

easily visible from telescopes on Earth. Yet although 'visible', they are so far impossible to recognize beyond doubt. A giant white hole will be indistinguishable from an exploding galaxy, and a relatively small one, of, say, ten Solar masses, will look at a great distance just like an ordinary star. J. V. Narlikar and his colleague K. M. V. Apparao, at the Tata Institute of Fundamental Research in Bombay, suggest that those very violently exploding galaxies known as Seyfert galaxies (after their original discoverer, Carl Seyfert) could be giant white holes, presumably of many millions of Solar masses, pouring back into the Universe matter which had been devoured by distant black holes.[11] They propose that the physical mechanism of a white hole will be identical to that of a black hole, except that everything will happen in reverse. Its behaviour, like the black hole's, will follow Roy Kerr's equations, and it will have a rate of spin in proportion to its mass. Now the nature of the Einstein–Rosen bridge, the region where forward distances are reduced to zero, predicts that the white hole will come into existence at the very same instant as its corresponding black hole. And so, if an astronaut can vanish down a rotating black hole without his ship being destroyed, then he will be able to emerge equally unscathed, a fraction of an instant later, from a rotating white hole. If the principle of symmetry is to be preserved, then a black hole and a white hole must represent the two opposite ends of an Einstein–Rosen bridge.

It may now be possible to guess the policy of that mysterious bureaucrat whom I have called Dr White. If instantaneous travel is to be achieved, starting from a sort of 'cosmic railway station' one light-year from the Sun, it will not be enough simply to construct a black hole at that point. The black hole would enable astronauts to travel instantaneously to distances of many light-years, but they would have no means of coming back, except through normal space, a journey which could take them many decades. It will be necessary *to construct a second black hole at the distant point where they will emerge into normal space, which they will use for return to a point no further than about one light-year from the Sun.*

This may sound a preposterous suggestion, but it will only be slightly more difficult than the construction of the first black hole. The difference will be that a second fleet of magnetic scoops, of the giant interstellar bull-dozers which I described in an earlier chapter, will have to be assembled after the first black hole has been constructed. Unlike the earlier scoops which will move into deep space fully operational, they will approach the completed black hole disassembled and compactly folded. All this, and what is to follow, will of course demand an extremely high level of machine technology. Yet judging by the rapid pace in the development of today's machine systems, there is every reason to suppose that the necessary technology will be available when it is needed. But why must the second fleet of magnetic scoops be 'compacted'? The answer lies in the mathematical prediction from Schwarzschild's equations that the width of the actual navigable gateway of a black hole of ten Solar masses will be little more than 600 yards. When the black hole is ready, it will be necessary to send these highly compacted, and obviously unmanned, machines into orbits round the spinning black hole that precisely match its own speed of rotation. They will then be transmitted, one by one, through the disc of the black hole, and out again into normal space through the white hole which will have come into existence at a point several light-years away.

Dr White faces here a problem somewhat similar to that which confronted the legendary Athenian hero Theseus after he had killed the Minotaur in the labyrinth at Knossos. The labyrinth, with its myriad stairways and passages, was said to have been complicated almost beyond human comprehension. Theseus could only find the monster's lair in its centre by letting himself be guided by the cadence of the creature's angry roars. Getting out again might have been doubly difficult, since he had no sounds to guide him. But he had marked his outward path with a long trail of string, given to him for that purpose by his mistress Ariadne. Following the string, he soon found the exit to the labyrinth.[12]

The task of the disassembled ramjet machines after they emerge from the white hole represents essentially the same problem as that which faced Theseus after his fight with the monster when he started to seek his way out. The ramjets will assemble themselves into operational form at the command of their computerized programmes. They will then take up their various positions and start bulldozing ten Solar masses of dust into one single point, in exactly the same way as the original ramjets behaved, under the direction of Dr Black. At last, after another period of at least fifteen years, in addition to the forty already elapsed, they will have constructed a second black hole. This second black hole, in turn, will bring into existence a *second white hole,* which will emerge in normal space at a point reasonably near to the Sun; and a safe distance would again be about one light-year. This may sound rather complicated, but see Figure 7—which is not of course drawn to scale.

The approximate positions of the four holes, two black and two white, is shown in the Figure.

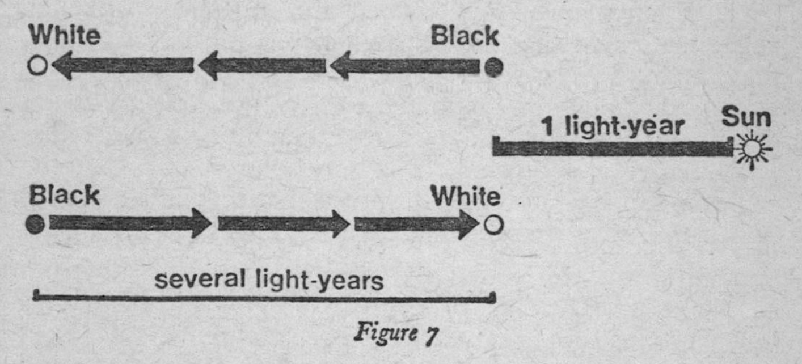

Figure 7

The arrows indicate the astronauts' directions of travel, moving in each case from a black hole to a white hole, the only direction in which travel is permitted.

From the view of an observer on the Earth's surface,

nothing very dramatic will seem to have happened when the white hole appears. A shining object, like a new star, will quietly appear in the night sky. In a sense, it will have 'exploded' into normal space, but this explosion, if the age-long leakage of ejected matter can be properly termed an explosion, will take millions of years. The white hole, being a mirror-image of a black hole, will take the same amount of time to eject matter as a black hole does to devour it.* No deadly emission of particles will reach the Earth or its neighbouring planets. Yet in technological terms something very dramatic will have happened. A complete circuit will have been made. Ariadne's string will have turned back upon itself; two parallel Einstein–Rosen bridges will have come into existence, with matter flowing through them in opposite directions, and each with a black hole and a white hole at their respective ends.

The place in deep space where the second black hole has been built will be the point, and at first the only point, to which astronauts from Earth will be able to make instantaneous journeys and return instantaneously to a point near home. But where will they have travelled to? To what part of the Galaxy will the first black hole lead? I have constantly evaded this question with the vague statement that their instantaneous journey will take them 'several light-years'. The answer is that, at present, we just do not know. Nobody, to my knowledge, has yet attempted any calculation to ascertain the probable distance between a black hole of ten Solar masses and its corresponding white hole. It is probable that this distance is in some way proportional to the masses of the holes,

*The rate at which a black hole devours matter (and by implication the rate at which a white hole ejects it) is predicted vividly by Roger Penrose in his contribution to Laurie John's *Cosmology Now*: 'Imagine having to drain a bath the size of Loch Lomond through a normal sized plug-hole.' If a white hole is the mirror-reverse of a black hole, no danger could therefore possibly arise from the appearance of a white hole of ten Solar masses at 6 trillion miles from the Sun.

but we have no idea yet how to express this relationship as an equation, and until we can do that there is no possibility of being able to write a navigational manual for instantaneous travel.

But today, in the last quarter of the twentieth century, 200 years or more before we can hope to put the plan into effect, this difficulty does not seem to matter very much. Judging from the rate of progress that has been made in mathematical physics during the last fifteen years, it cannot be long before it is solved.

But it can be seen, none the less, that the plan will work in principle. For the two Einstein–Rosen bridges connecting the two pairs of holes will be parallel, as Figure 7 showed. And being parallel, and resulting from two black holes of equal mass, the distances in normal space covered by the two bridges will be equal. And so astronauts will always be able to return home.

Setting out from the Earth, they will of course have to travel a whole light-year, some 6 trillion miles, in normal space, before they can begin an instantaneous journey. And so when we call the entire journey 'instantaneous', we are not speaking literally. The distance of one light-year between the Earth and the two holes will mean that in effect the astronauts will be away from home for at least several years. Yet what a scientific and technological triumph their journey will have been! They will have traversed a part of the Galaxy in a time no greater than it takes us to snap our fingers.*

But a single black hole–white hole parallel system will be of no more use to Galactic travellers than a bus

*What will it cost to set up this cosmic transport system. A very rough guess, at 1974 prices, might be about $3 trillion. This is 120 times the cost of the Apollo moon-landing program, and about two-thirds of the Gross World Product in 1974. But if we assume an average annual growth in real terms of 3 per cent in the human economy up to the beginning of the twenty-third century (a very reasonable extrapolation from present long-term trends) the cost would be a mere thousandth of the G.W.P., a price which humanity could easily afford.

route would be today that went to one place only and permitted no stops along the way, or a city telephone system in which only two citizens had telephones. Innumerable other such systems will eventually have to be built before a network of instantaneous routes can exist between the planetary systems of suitable neighbouring stars. Even after that, before it will be possible to explore and colonize our entire Galaxy of 100,000 million suns, there will have to be an era of hole construction that lasts for many thousands of years.

Those with any sort of technical background will have perceived many objections to this general plan. Some can be answered easily, but other represent problems whose solutions must await the engineering skills of future generations. How, it may be asked, can we believe in the competence of a fleet of machines to build a second black hole, operating at a place so far away that there will be no possibility of human supervision and control? And how can we be sure that the second white hole will not emerge in the wrong place? Instead of harmlessly coming into existence 6 trillion miles away, might it not, through some mechanical misjudgement, explode at the very centre of our Solar System, and destroy all human life?

Even today, one can see that this problem is not insoluble. We entrust our lives to a machine intelligence each time we enter an automatic lift. Our local planets are continually visited by unmanned spacecraft whose manoeuvres, controlled only by radio messages from distances of many millions of miles, have a sophistication that even thirty years ago, would have been considered the wildest science fiction. Following a prearranged programme, they change their shapes, then float down by parachute through the atmospheres of alien planets after pulling their own rip-cords, and from deep inside their innards they unfold mechanical legs for soft landings. Yet I am not talking of what machines can do today. I am talking of what they will be able to do in perhaps two and a half centuries from now! On this time-scale of machine evolution, the objection appears absurd. It is surely realistic to predict, for that age, the feasibility of safety systems which will make disasters not only highly improbable but physically impossible.

Yet the new system will bring about one minor change which will be truly irreversible. The Sun will no longer orbit the Galaxy as a single, sovereign star. The presence of a black hole and a white hole so near to it will change the local gravitational fields to the extent that the Sun will become a three-star system, and the effect on its orbit round the Galaxy will be noticeable on Earth. The Sun will *move* in relation to other stars in ways that it would not otherwise have moved. It will not matter greatly. It will affect only the lives of one of the smallest minorities of people, those people who have the task of publishing atlases of the stars, and of writing next year's *Astronomical Almanack,* which tell us that such-and-such a star will appear at such-and-such an hour at so-many degrees above one's back garden. For all this information will become slightly wrong! A particular star or constellation may reach its zenith an hour later than predicted. Many learned books will have to be rewritten. It is even conceivable that audiences at productions of Shakespeare's *Julius Caesar* will titter when the dictator declaims:

> I am constant as the northern star,
> Of whose true-fixed and resting quality
> There is no fellow in the firmament.

For the northern star, Polaris, will no longer be 'constant' or 'true-fixed', in the place where Shakespeare's contemporaries and we today see it. It will have moved —or rather we shall have moved in relation to it. The machines at all the world's planetariums will have to be reprogrammed, since the information which they project on to their domed ceilings will no longer be accurate. The planets will still appear in in the same places, as seen from Earth, since they move with the Sun. But the stars, being much more distant, and being relatively unaffected by the Sun's gravitational field, will be in slightly different places.* Yet this minor inconvenience, however tedious it may be for the bureaucrats of astronomy, will be a small price to pay for the boon of instantaneous interstellar flight.

*I must admit that I have not taken the trouble to calculate the new positions of the stars and constellations. The task should represent an interesting mathematical problem.

Epilogue

I have now reached the end of this account of how instantaneous flight in space can be achieved. It has been in many ways crude and imperfect, since, wishing only to establish certain fundamental principles, and remove what might have seemed basic objections, I have deliberately ignored many minor but important problems. Some technical questions are discussed in the Appendices, but others must await the more advanced technologies of future ages. How, for example, will the bodies of astronauts be shielded from the tremendously powerful X-ray emissions that are found near the discs of black holes? Must their ships have outer hulls of thick lead, even more massive than the walls of a modern nuclear power station? This safety requirement could magnify sixfold the amount of energy required for their acceleration to speeds of 400 million miles per hour as they match velocities with the spinning disc of the black hole. Or will the metallurgists and chemists of that era have discovered some much lighter substance, which will not add greatly to the masses of the ships, but will deflect from their hulls all dangerous energetic particles, as a mirror reflects back light-rays? One suspects that they will, since metallurgy and chemistry, like most other sciences, are today in their infancy.

Perhaps also, as astronauts enter the disc aperture, they will experience the pulling effects of a mighty tide.

By choosing their route with the cunning that I have described in Chapter 3 they will have avoided the crushing densities of the central singularity, but they will nevertheless be exploiting the gravitational energy of ten Suns, something more than a luckless swimmer on Earth feels when he is pulled away from the beach by an ebbing tide! Who knows what devices will be needed to counteract this tide, to prevent it from twisting the bodies of astronauts into something which no biologist could recognize as ever having been human. We can also predict another natural safeguard against tidal crushing: because the spaceship will be moving instantaneously, the tide will not pull upon the astronauts for more than perhaps a millionth of a second. In these conditions, it is possible that they may be unaffected, because the atoms which comprise their bodies and brains, and the spaceship itself, will have had *no time in which to respond to the pull of the tide*. In short, they will pass through the inner event horizon before the gravitational tide has had time to affect them, like the circus animal which leaps fast enough through the ring of fire and is not burned. Yet the very grandparents of the engineer whose life's work will be to solve the details of these problems have probably not yet been born.

If the bodies of the travellers can be protected, what then of their minds? When three men were killed in 1881 in the gunfight at the O.K. Corral in Tombstone, Arizona, that vigorous newspaper the *Tombstone Epitaph* was not content merely to record their deaths. It stated that they had been 'hurled into eternity in the duration of a moment'. How will it feel to be hurled several light-years and survive, all in the duration of a moment? Professor John Taylor, of London University, discussing recently in a radio broadcast a journey through a black hole, suggested that the astronauts might have severe psychological problems.[1] One can well believe it. And yet no psychological problems have ever by themselves prevented people from doing what they wanted to do. A person may be driven mad by fear of imminent personal destruction. But this is a physical fear, based on reasoned calculation. Someone of delicate imagination might indeed panic at the sheer awesomeness of being hurled across light-years, but neither 'awesomeness' nor any of its side-effects, whatever they may be, are likely to imperil healthy people who know the hazards of what they are doing, and have decided that those hazards are insignificant.

If a black hole transport system has natural safeguards against accidents, it is highly vulnerable to human malice. The navigable entrance to a black hole, that 640-yard aperture, could very easily be wrecked by military enemies or even by well-equipped terrorists. Professors Michael Simpson and Roger Penrose showed in 1973 that the inner event horizon of a black hole is likely to be unstable, in the sense that a sufficiently large physical disturbance to it will cause it to 'jam' (my word for it) and turn into a simple crushing singularity.[2] The black hole would then become impenetrable to spaceships. This act of sabotage could be carried out by detonating a nuclear bomb inside the aperture. The most frightening aspect of such an event would be that, as far as we can tell today, the black hole would be impossible to repair. There would be no way to approach it sufficiently closely to exert sufficient force to 'unjam' it. It would be therefore necessary to expend another forty years on building a fresh black hole. A planetary system could be cut off for that period from all traffic with its neighbours. This would be a serious matter in a future age when people on hundreds of thousands of Galactic planets are likely to be living from interstellar trade. Ingenious security methods may be needed to protect Galactic society from the schemings of the human military mind.

One's final reflection is that the whole enterprise, of building black holes and white holes, and flying through them like bees flying in and out of a hive, may in the end prove unnecessary. I do not mean that mankind will give up its dream of flying to the stars. That decision would require a fundamental change in human nature, of which there is little likelihood. But it may be, and the chances of this appear very remote, that some of Einstein's equations may have been in error, and that it may, after all, be possible to fly through space faster than light in defiance of his Special Theory. Some writers believe this emphatically, but so strong is the case against them, both from theory and experiment, that one cannot read their works without laughing.

Yet one can see easily that there is an emotional as well as a scientific basis for challenging the laws of physics as they are known today. Intellectual establishments are often suspected of complacency, and indeed are often guilty of it. So much knowledge has now been acquired that it has become very easy to imagine that everything important has now been discovered. Many astronomers talk today with the easy assurance of people who think they know everything. One might suppose from listening to them that they have personally visited every star in the Galaxy, and are in the habit of composing unified field theories as a warm-up before breakfast. In a lecture in 1975, the Russian astronomer V. L. Ginzburg pointed out the dangers of this attitude:

> In the past, fundamental physical theory was several times thought of as completed, and then that was found not at all true. Therefore, there seems no doubt of the necessity to overstep the limits of conventional physics. And if that is so, 'new physics' should, of course, be sought. Those who avoid it condemn themselves to failure in discovering anything really important.[3]

It is conceivable, therefore, that "new physics' may overrule some of Einstein's equations just as Einstein's own discoveries overruled much of the earlier work of Isaac Newton. If this happened, it might be found that the speed of light represented no barrier to acceleration in normal space, and that astronauts could travel at any speed they wished. Conceivable it may be, but it is very unlikely. I have written this book on the assumption that Einstein's equations, and Roy Kerr's solution to them, represent a correct view of nature, and that black holes will need to be constructed for the achievement of interstellar travel.

I hope only that I have succeeded in burying a myth, the myth that the barrier of the speed of light will be for ever impassable. There may be something brutal, something vaguely offensive to our ancient desire for cosmic serenity, about the method I have proposed to overcome it. Yet sophisticated refinements to it can no doubt be made. But there is one great argument in its favour: it will work. Those daunted by its great technical difficulties should take courage from the last sentence of Robert Bussard's famous paper of 1960, in which he outlined his plan to build a ramjet: *'Nothing worthwhile is ever achieved easily.'*[4] Moreover, as Dr Black correctly pointed out, it would not endanger the Earth or any other future interplanetary stronghold of mankind. The myth, I now believe, is at last buried. It will not be easily exhumed.

Appendices

Appendix I

Diameters of the Navigable Apertures of Black Holes of Different Masses

To calculate the approximate diameter of a black hole, we multiply the amount of mass from which it has been formed by the formula 4 G/c^2, where G is the gravitational constant $6{\cdot}67 \times 10^{-8}$, and c^2 is the square of the speed of light; c^2 works out conveniently, using centimetres per second, as 9×10^{20}.

Suppose that we want to know what the diameter of the Earth would be if it were miraculously compressed into a black hole. (I say 'miraculously' since it is unlikely that natural force could compress an object into a black hole unless it is at least three times the mass of the Sun; its gravity will simply not be strong enough.) The Earth's mass is 6×10^{27} grams (6×10^{21} tons). Multiply this by $4G/c^2$, and while the Earth will have lost none of its mass, its diameter will have shrunk to a mere 1·8 centimetres, or about two-thirds of an inch!

The 'navigable' aperture of such a tiny black hole, the height of its spinning disc-edge, would be no more than a hundredth of this; plainly a very small spacecraft, with microscopic crew members, would be needed to get through it. This gives some idea why a large amount of mass is needed for a useful black hole. Table 2 gives some figures of black holes of different masses, with their diameters and the diameters of their navigable apertures. I should emphasize again that the data in the first two lines is imaginary, since they describe insufficient mass for a black hole. For convenience, I give equivalent figures in miles and feet.

Table 2

Solar masses (Sun = 1)	*Diameter of black hole (km)*	*(Miles)*	*Navigable aperture diameter (Metres)*	*(Feet)*
1	5.9	(3.7)	60	(198)
2	11.9	(7.4)	120	(393)
3	17.8	(11.0)	180	(591)
4	23.7	(14.7)	240	(786)
5	29.6	(18.4)	300	(984)
6	35.6	(22.1)	360	(1,182)
7	41.5	(25.7)	420	(1,378)
8	47.4	(29.4)	470	(1,542)
9	53.4	(33.1)	530	(1,740)
10	59.3	(36.8)	590	(1,935)

Black holes of much greater mass would of course have wider navigable apertures. A black hole of a million Solar masses for example would have a diameter of 5·9 million kilometres (3·7 million miles) and a navigable aperture of 59,000 kilometres (37,000 miles). But the construction of black holes very much larger than ten Solar masses would be very difficult since the raw material does not appear to exist at a reasonable distance from the Sun. There would also be an increasing risk of disruption of the Solar System by the black hole's gravitational field. Ten Solar masses seems a fair compromise, both on grounds of feasibility and safety.

Appendix II
Some Milestones in the Development of Black Hole Theory

1798 Peter Simon Laplace suggests that some stars may be so massive that their gravity will prevent light from escaping fom them.

1905 Einstein publishes his Special Theory of Relativity.

1916 Einstein publishes his General Theory of Relativity. Karl Schwarzschild carried Einstein's equations to their most extreme conclusion, and suggests the possibility of black holes which do not rotate. For the first time, he introduces the idea of a 'singularity' in space-time.

1935 Einstein, with Nathan Rosen, suggests that separate parts of normal space may be interconnected by timeless 'bridges'.

1939 J. Robert Oppenheimer and Hartland Snyder analyse in detail for the first time the mechanics of total gravitational collapse.

1960 M. D. Kruskal draws the first 'Kruskal diagram', showing the behaviour of space and time in the neighborhood of a Schwarzschild, non-rotating black hole.

1963 Roy Kerr incorporates fast rotations into Schwarzschild's model of a black hole. He produces the so-called 'Kerr solution' to Einstein's 1916 equations.

1967 R. H. Boyer and R. W. Lindquist draw a Kruskal diagram for a rotating black hole, indicating that an instantaneous journey could be made through it which did not hit the singularity.

1975 David Robinson shows that all black holes must rotate, and that Kerr's solution must be correct.

Appendix III
A Do-it-yourself Guide to Einstein's Two Main Theories of Relativity

1 The Special Theory of 1905

The Special Theory rests on the still unexplained fact that the speed of light in the vacuum of space is constant at 670 million m.p.h. (or 186,000 miles per second, or 300,000 kilometres per second), *irrespective of the speed of its source*. I have explained in the Introduction how it must then follow that the *length* of a moving spacecraft in the direction of motion must be reduced in inverse proportion to its speed. The spacecraft's *mass,* i.e. the energy required to accelerate it, must also increase as the object accelerates. And all clocks aboard the spacecraft—the term 'clock' here includes both the astronaut's wristwatch and the ageing process of his body—must slow down as the spacecraft accelerates.

The extent of these changes is calculated by three simple formulae. To calculate the *length* of a moving spacecraft, we multiply its length at rest, that is its length when stationary, by the formula

$$\sqrt{\left(1-\frac{v^2}{c^2}\right)}$$

where v is the speed of the spacecraft, and c is the speed of light. The increased mass of the moving spacecraft is calculated by a slightly different formula. Its mass at rest is multiplied by

$$\frac{1}{\sqrt{\left(1-\frac{v^2}{c}\right)}}$$

Measuring the slowing of time on board the spacecraft is equally simple. To see how much more slowly the astronauts are ageing, we multiply a given period, say 60 minutes of earth-time by

$$\sqrt{\left(1-\frac{v^2}{c^2}\right)}$$

Two spaceships cannot recede from one another at a combined speed greater than that of light. If a man standing on the ground sees one craft racing overhead going due north at 90 per cent of the speed of light, and another going due south at the same speed, he might suppose that each craft was receding from the other at a combined speed of 180 per cent of that of light. Yet he would be wrong. The sum of the two speeds cannot exceed that of light. It must be calculated by the formula

$$\frac{a+b}{1+\frac{ab}{c^2}}$$

where a and b are the respective speeds of the two craft, and c is the speed of light. It will be seen from this formula that if the two craft had been travelling very slowly, say at 600 m.p.h., then the sum of their speeds would be about 1,199·9999999 m.p.h., or *almost* 1,200 m.p.h. But if their speeds are very great, the formula gives quite a different kind of answer. Pretend that instead of two spaceships, the man sees two light-beams receding in opposite directions. He will estimate their combined speed of recession as twice the speed of light, or 2c. But if he were riding on one of the beams, he would estimate the other's speed of recession, according to the formula, as

$$\frac{c + c}{1 + \frac{c^2}{c^2}}$$

which of course works out as c.

Let us therefore test an imaginary spacecraft at varying speeds, and see how its length, ageing and mass change according to Einstein's equations. We will assume that a ship, while stationary in port, is exactly 100 metres in length, and has a rest-mass of 100 tons. As the ship accelerates we see that a ship-hour becomes a progressively smaller fraction of an earth-hour.

Table 3

Speed of ship as percentage of light	*Length of ship (metres)*	*Mass of ship (tons)*	*Duration of ship-hour in minutes (Earth = 100)*
0	100·00	100·00	60·00
10	99·50	100·50	59·52
20	97·98	102·10	58·70
30	95·39	104·83	57·20
40	91·65	109·11	55·00
50	86·60	115·47	52·10
60	80·00	125·00	48·00
70	71·41	140·03	42·75
80	60·00	166·67	36·00
90	43·59	229·42	26·18
95	31·22	320·26	18·71
99	14·11	708·88	8·53
99·9	4·47	2,236·63	2·78
99·997	0·71	14,142·20	1·17
100	zero	infinity	zero

As the figures in Table 3 plainly indicate, no spacecraft can ever travel at the speed of light itself. An infinite mass would require an engine of infinite power for its propulsion. And even if this were miraculously achieved, the spacecraft's length would be zero, and it could not therefore exist.

2 The General Theory of 1916

The General Theory is more subtle than the Special, although the argument is, in my opinion, if anything easier to follow. But its conclusions are so strange that many people have found them very difficult to grasp.

This theory follows directly from Einstein's realization that all the phenomena of the Special Theory, the slowing of time in a fast-moving vehicle, that vehicle's contraction of length in the direction of motion, its increase in mass, have a single fundamental cause—the effects of *gravity*.

In essence, Einstein's reasoning was as follows: whenever a vehicle, let us say a spaceship, accelerates, it is using its energy *against* the gravitational field which would otherwise have kept it in its orbit. Yet even though it is now moving in a different direction, all it has really done is to move from a smaller orbit to a much greater one. Whereas hitherto, let us say, it was in orbit round the Earth, it is now in a much larger orbit round the Sun.

Suppose that it now accelerates a second time and moves out of the Sun's immediate gravitational field. It is *still* in orbit, this time around part of the Galaxy. However many times that ship accelerates, it will always find itself in some new orbit around a mass. And the path of the orbit will always be, if only very slightly, curved.

This is so because of the very structure of space and time which exist, and only exist, *because of the presence of matter*. Matter creates the space which it occupies. If there were no matter in the Universe, neither stars nor galaxies, then not even the spaces between the stars and galaxies would exist! To state this strange idea bluntly, *nothing whatever would exist*. And there would be no such thing as time.

Space and time must therefore be considered as physical entities which are curved by the presence of a large mass. The larger the mass, the more extreme is the curvature. In the case of a black hole, which is so massive that the density of matter becomes infinite, the curvature of space and time around it also becomes infinite. We have then a 'singularity', a region where the conventional laws of physics break down completely.

The actual curvature of space and time in the presence of a large mass can be calculated by the formula

$$10\sqrt{\frac{GM}{1{,}000\ rc^2}}$$

where G is the gravitational constant, $6{\cdot}67 \times 10^{-8}$, M is the mass of the object, and r is its radius. As Eddington discovered in his famous experiment in 1919, the curvature of space and time in the presence of the Sun is approximately 1·7 seconds of arc. (The formula above yields this result if we remember to multiply the answer by 3,600, since that is the number of arc seconds to a degree. The Sun's radius should be taken as 7×10^{10} centimetres, and its mass as 2×10^{33} grams, and the speed of light, in centimetres per second, is 3×10^{10}.) To obtain his result, Eddington measured the position of certain stars at the Sun's rim during the darkness of a solar eclipse. Their normal positions were displaced by about 1·7 arc seconds, thus proving the curvature of light rays, and hence also of space and time. In the case of a shrinking black hole, where we have a much greater mass and a very small radius, the curvature soon exceeds 360 degrees. In other words, it becomes infinite.

Both theories, the Special and the General, have been repeatedly tested and confirmed by experiment.

Glossary

1 Mathematical notation

In some of the following notes, as in the text, it has been necessary to write some very large and some very small numbers. The mass of the Sun, for instance, is about 2,000,000,000,000,000,000,000,000,000 tons, and the average density of the Orion Arm is about 0·000000000-0000000001 grams per cubic centimetre. It is plainly absurd to write numbers in this way, since we would have the tedium of counting the zeros and continually double-checking for errors. Instead, it is easier to write the Sun's mass as 2×10^{27} tons, and the density of the Orion Arm as 1^{-19} grams per cubic centimetre. 2×10^{27} means simply 2 followed by 27 zeros, and 10^{-19} means a decimal point followed by 18 zeros and then a 1.

Why only 18 zeros and not 19? It is because the 'exponential' system of powers of 10 goes like this:

1,000,000 = 10^{6}
100,000 = 10^{5}
10,000 = 10^{4}
1,000 = 10^{3}
100 = 10^{2}
10 = 10^{1}
1 = 10^{0}
0·1 = 10^{-1}
0·01 = 10^{-2}
0·001 = 10^{-3}
0·0001 = 10^{-4}
0·00001 = 10^{-5}
0·000001 = 10^{-6}

And so on. Suppose that we wish to multiply 400,000,-000 by 2,000,000,000. The exponential system makes it very easy. The sum is,

$$(4 \times 10^8) \times (2 \times 10^9)$$

We add together the exponents 8 and 9, making $4 \times 2 \times 10^{17}$, or 8×10^{17}. To divide 400 million into 2,000 million is just as easy. 8 is subtracted from 9, leaving 1. This gives $2 \div 4 \times 10^1$ which works out as 5.

Throughout the following References and Notes (as opposed to the text, where people seem to prefer the old imperial measurements), I have given all densities in grams per cubic centimetre, distances in centimetres or light-years, and masses in grams. I have put in brackets the equivalent amounts in pounds per cubic inch, statute miles and tons. To convert grams per cubic centimetre to pounds per cubic inch, multiply by 0.0361. To convert back again, multiply by 27·68. There are $9{\cdot}5 \times 10^{17}$ centimetres in a light-year ($5{\cdot}9 \times 10^{12}$ miles), and 1 million grams make a ton.

2 Words and phrases

The following may assist some readers, even if many of the terms are familiar to others.

Asteroid A tiny planet, usually less than 400 miles in diameter. There are estimated to be between 50,000 and 100,000 asteroids in the Solar System.

Black hole A region in space where matter vanishes from this part of the Universe, connecting to a white hole (q.v.) in another part.

Bussard ramjet A future space vehicle, first proposed by Robert Bussard in 1960, which would deploy giant magnetic fields to suck in interstellar hydrogen for fuel. A version of this system, as proposed in this book, would 'bulldoze' great quantities of interstellar iron dust and hydrogen plasma (q.v.).

Causality, Principle of The principle which states that a consequence cannot occur before the event that caused it.

Any theory of the Universe which involves a violation of this principle must be invalid.

Constellation Any one of 88 areas of the sky, arbitrarily mapped out for astronomical convenience.

Density The average amount of matter which is concentrated in a given space. It is usually computed in grams per cubic centimetre or pounds per cubic inch. Table 4 lists some typical densities, from the lowest to the highest. I give them both in grams per cubic centimetre and pounds per cubic inch. All must be regarded as approximate.

Table 4

	gm./cm.^{3}	*lbs/in.*^{3}
Intergalactic space	4×10^{-31}	$1{\cdot}4 \times 10^{-32}$
Interstellar space	10^{-27}	$3{\cdot}6 \times 10^{-29}$
The Orion Arm of the Galaxy	10^{-19}	$3{\cdot}6 \times 10^{-21}$
Terrestrial air at sea level	$1{\cdot}3 \times 10^{-3}$	$4{\cdot}7 \times 10^{-5}$
Pure water	1	0·0361
Average density of the Sun	1·41	0·05
Average density of the Earth	5·42	0·2
Lead	11·34	0·41
Interior of a white dwarf	3×10^{5}	$1{\cdot}1 \times 10^{4}$
Interior of a pulsar	$6{\cdot}2 \times 10^{12}$	$2{\cdot}2 \times 10^{11}$
Average density of a black hole	8×10^{15}	3×10^{14}
Singularity of a black hole	more than 10^{94}	more than 4×10^{92}

Differential rotation A characteristic of a system, such as a galaxy or a black hole, whose outer regions rotate at different speeds from the inner ones.

Einstein–Rosen bridge A passage directly connecting one part of the Universe to another, such as a black hole–white hole system. Such 'bridges' were proposed in 1935 by Einstein and Nathan Rosen, and confirmed by other theorists after Kerr published his solution to Einstein's equations in 1963.

Electron One of the fundamental constituents of matter, which orbits the nucleus of every atom.

Escape velocity The speed which an object needs to escape from a planet's (or star's) gravitation into space. The Earth's escape velocity is 11·2 kilometres per second (25,000 m.p.h.). One useful definition of a black hole is of a celestial body whose escape velocity is greater than the speed of light, i.e. more than 300,000 kilometres per second (670 million m.p.h.). A body's escape velocity can be calculated from the formula

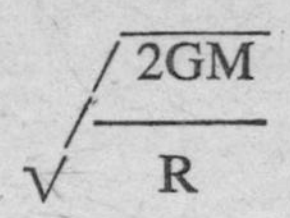

$$\sqrt{\frac{2GM}{R}}$$

where G is the gravitational constant of $6{\cdot}67 \times 10^{-8}$, M is the object's mass in grams, and R is its radius in centimetres.

Event horizon The surrounding region of a black hole, beyond which it is impossible to communicate with the outside Universe, because the escape velocity would be faster than light. A rotating black hole has two event horizons, an outer and an inner. A black hole–white hole system (or an Einstein–Rosen bridge) would therefore have four event horizons, two inner and two outer.

Ferromagnetism The normal phenomenon of magnetic attraction. The metals most strongly attracted by a magnet are iron, nickel, cobalt and their alloys.

Galaxy A large group of stars. The galaxy to which our Sun belongs contains about 180,000 million stars. It is about 100,000 light-years (q.v.) from end to end. There are tens of thousands of millions of galaxies in the Universe.

Implosion The opposite of an explosion; everything is crushed inwards instead of being hurled outwards.

Intergalactic space The space between the galaxies.

Interplanetary space The space between the planets of our Solar System (q.v.) or any other solar system.

Interstellar space The space between stars.

Ionize Literally, to separate electrons from atoms. In the context of this book, to turn clouds of inert interstellar hydrogen into plasma (q.v.) which will then be magnetic.

Kerr solution The solution to Einstein's 1916 equations published by Roy Kerr in 1963, showing that black holes should be considered as rapidly-spinning objects. This contrasted with the older Schwarzschild solution (q.v.) which assumed that black holes were stationary on their axes. Recent work has shown that the Kerr solution must be correct, and that a black hole must be the opening to an Einstein–Rosen bridge (q.v.)
Kruskal diagram A map or a diagram of space and time inside a black hole.
Laser beam A beam of extremely intense light which can produce great heat at the point of impact. New techniques of increasing the power of laser beams are continuously being discovered.
Light-year A common unit of distance in astronomy. It is the distance that light travels in one year, going at 1,000 million kiometres per hour, or 670 million m.p.h., or 186,000 miles per second. It is thus about 9•5 trillion (q.v.) kilometres, or 5•9 trillion statute miles.
Neutron star See Pulsar.
Normal Space Ordinary interstellar space.
Photon A particle of light. Light from a star reaches us in the form of immense numbers of photon particles.
Plasma The fourth state of matter, beyond the three states of solid, liquid and gas. It is matter whose electrons have been stripped away from the atoms. All substances become plasma when heated to about 22,000° Centigrade, 40,000° Fahrenheit.
Pulsar, or neutron star. The remains of any star of between about 1½ and 3 Solar masses (q.v.). A pulsar is only a few miles across, but with a density (see above) of about 6 trillion grams per cubic centimetre, or 100 million tons per cubic inch.
Quasars Mysterious objects, apparently at the very edge of the Universe, which are estimated to be shining each with the brilliance of 100 galaxies. Some physicists believe that quasars may be giant white holes (q.v.).
Relativity, Theories of The two great theories of Albert Einstein. The Special Theory of 1905 showed that no object can reach or exceed the speed of light. However,

time inside a moving vehicle slows down in inverse proportion to its speed. The General Theory of 1916 describes the actual curvature of 'warping' of space in the neighbourhood of a massive object. For a full explanation of the two main theories of relativity, please see Appendix III.

Schwarzschild solution The first mathematical description of a black hole, depicting a stationary, non-rotating black hole, that Karl Schwarzschild formulated in 1916 as a 'solution' to Einstein's General Theory. It is now known to be incorrect in several respects. But Schwarzschild's formula for calculating the diameter of a black hole from its original mass is still valid and is an essential tool. (See Note 5 of Chapter 3).

Singularity The point at the centre of a black hole where density is almost infinite, and where everything which enters it is crushed out of existence.

Solar mass Any mass equal to the mass of the Sun, which is 2×10^{33} grams (2×10^{27} tons). It is often convenient to describe very large objects in Solar masses rather than in grams or tons.

Solar system or stellar system. The region of space occupied by a sun and its planets.

Spiral arm A great arm consisting of relatively condensed matter in the form of gas, dust, and stars, that twists out from the centre of the Galaxy. The Orion Arm is one of these.

Star A sun, of which the nearest to our own is more than four light-years distant.

Supernova The total disruption of a star by violent explosion.

Symmetry, Principle of The principle which states that all forms of matter and all phenomena must have their opposites. Thus, there are matter and anti-matter, black holes and white holes.

Thought experiment A fictitious story used to illustrate a scientific principle. To make their points, they must sometimes have the most improbable details, such as infallible fortune-tellers.

Trillion A million million; 1 followed by 12 zeros.

White dwarf The final form of the burned-out remains of a star of up to about 1½ Solar masses. It is very dense, although less dense than a pulsar (q.v.).
White hole The opposite of a black hole, in which matter is ejected instead of being sucked in. Its dynamics are identical to those of a black hole—except that everything occurs in reverse. A black hole and a white hole are believed to form the two ends of an Einstein–Rosen bridge.

References and Notes

Introduction

1 G. K. O'Neill, 'The Colonisation of Space', *Physics Today*, September 1974; also, A. Berry, 'Pioneers Who Aim to Live in the Sky', London *Sunday Telegraph*, June 29th, 1975; E. Burgess, 'Colonies in Space', *Astronomy*, January 1976.
2 Adrian Berry, *The Next Ten Thousand Years*.
3 *The World Almanack and Book of Facts*, 1976 and preceding issues. The Gross World Product works out as being about 5 times the U.S.A.'s G.N.P. These figures were taken from statistics of the Agency for International Development.
4 See, for example, *Main Economic Indicators*, September 1975, pp. 148–9, published by the Organization for Economic Cooperation and Development (O.E.C.D.).

Chapter 1: *Where Matter Vanishes*

1 See for example Dole, *Habitable Planets for Man*, p. 103. It appears likely that most stars have planets, although the great distances of interstellar space make their detection from Earth very difficult and their actual observation so far impossible. A Jupiter-sized planet has been detected in orbit around Barnard's Star, six light-years from Earth. It has even been possible to observe planets during their age-long process of being formed. See G. F. Gahm *et al.*, 'The T Tauri star RV Lupi and its Circumstellar Surroundings', *Icarus*, vol. 24, pp. 372–8, 1975. Also, *Journal of the British Interplanetary Society*, vol. 28, no. 7, July 1975, p. 493.

2 Ptolemy's *Guide to Geography.*
3 In this experiment, conducted in 1972, two American physicists, Joseph Hafele and Richard Keating, flew round the world from west to east in a commercial jet, carrying with them an atomic clock of great accuracy to which they had synchronized another such clock on the ground. On their return, the airborne clock was found to be running about 89 billionths of a second behind the ground-based clock, thus confirming Einstein's prediction. See my account of this experiment. 'Man Who was Younger than His Own Children', London *Daily Telegraph,* October 29th, 1972.
4 Dover Publications of New York have done great service to scientific understanding by republishing in 1952 Einstein's 1905 and 1916 papers in English, together with important papers by Lorentz, Weyl and Minkowski, under the title, Einstein, *The Principle of Relativity.* The original Einstein papers, are (Special Theory): A. Einstein, 'Zur Elektrodynamik bewegter Korper' ['On the Electrodynamics of Moving Bodies'], *Annalen der Physik,* vol. 17, 1905; (General Theory): A. Einstein, 'Die Grundlage der allgemeinen Relativitätstheorie' ['The Foundation of the General Theory of Relativity'], *ibid.,* vol. 49, 1916. Also, A. Einstein, 'Kosmologische Betrachtungen zur allgemeinen Relativitätstheorie', ['Cosmological Considerations on the General Theory of Relativity'], *Sitzungsberichte der Preussichen Akad. d. Wissenschaften,* 1917.
5 From the famous first paragraph of a famous paper: J. A. Wheeler and R. W. Fuller, 'Causality and Multiply Connected Space-Time', *Physical Review,* vol. 128, no. 2, October 15th, 1962, pp. 919–29.
6 K. S. Thorne, 'The Search for Black Holes', *Scientific American,* December 1974, vol. 231, pp. 32–43.
7 Arthur Eddington, *Stars and Atoms,* p. 50; this passage is quoted in Walter Sullivan's excellent article on black holes, 'Curiouser and Curiouser: A Hole in the Sky', *New York Times Magazine,* section 6, July 14th, 1974.
8 From Penrose's contribution to Laurie John's *Cosmology Now,* pp. 103–26.
9 F. J. Dyson, 'Old and New Fashions in Field Theory', *Physics Today,* June 1965, vol. 18. A more detailed ac-

count of the evolution of the General Theory, and an account of the last weeks of Schwarzschild's life is given by S. Chandrasekhar, 'Development of General Relativity', *Nature*, vol. 252 November 1st, 1974, pp. 15–17.

10 From Brandon Carter's introduction to the Institution of Electrical Engineers' *Black Holes 1970–74*.

11 R. P. Kerr, 'Gravitational Field of a Spinning Mass as an Example of Algebraically Special Metrics', *Physical Review Letters*, vol. 11, no. 5, September 1st, 1963, pp. 237–8.

Chapter 2: *The Spinning Gateway*

1 It is now accepted that during the 'life-time' of a black hole—i.e., when it is devouring matter—it must rotate in accordance with the Kerr solution. Two papers appear to have proved this to the satisfaction of most astrophysicists. See, R. Wald, 'On the Uniqueness of the Kerr-Newman Black Holes', *Journal of Mathematical Physics*, vol. 13, no. 4, April 1972, pp. 490–9; and, D. C. Robinson, 'Uniqueness of the Kerr Black Hole', *Physical Review Letters*, vol. 34, no. 14, April 7th, 1975, pp. 905–6.

2 Robinson, op. cit.

3 I have based my invention of Dr Malevolent on a real accident. A technician at the British Aircraft Corporation was experimenting one day with a spinning fly-wheel. There was a defect in its material, and the entire wheel suddenly fragmented. The largest piece flew past his head, narrowly missing it, at several hundred miles per hour. It smashed its way through the wall of the laboratory, and was later found in a field three-quarters of a mile away. It can often be extremely dangerous to stand close to any rapidly spinning object, especially if one is standing parallel to the spinning axis.

4 See, for example, K. S. Thorne, 'The Search for Black Holes', *Scientific American*, December 1974, vol. 231, pp. 32–43. Also, P. C. Peters, 'Black Holes: New Horizons in Gravitational Theory', *American Scientist*, vol. 62, September–October 1974, pp. 575–83.

5 I have assumed for convenience in this case that the black hole is ten times the mass of the Sun. The formula for calculating the density of its equatorial strip will be,

$$\frac{1}{(\text{Number of rotations} \times 2\pi)^2 \times \text{radius}}$$

Taking the number of rotations per second as 1,000, π as 3•142, and the radius of the black hole as 29•6 kilometres 18 miles), this density will be $8{\cdot}5 \times 10^{-15}$ grams per cubic centimetre (3×10^{-16} pounds per cubic inch.) The average density of the black hole of ten Solar masses will be about 8×10^{15} grams per cubic centimetre ($2{\cdot}9 \times 10^{14}$ pounds per cubic inch). If we divide 8×10^{15} by $8{\cdot}5 \times 10^{-15}$ we obtain (approximately) 10^{30}.

6 For lucid accounts of *why* a black hole must have this shape, see the articles by Thorne and Peters cited in Note 4 above.

7 M. D. Kruskal, 'Maximal Extension of Schwarzschild Metric', *Physical Review,* vol. 119, no. 5, September 1st, 1960, pp. 1743–5. Kruskal diagrams are lucidly explained for the layman in Chapter 7 of William J. Kaufmann's *Relativity and Cosmology.*

8 The horizontal Kruskal diagram of the Kerr solution was first formulated by R. H. Boyer and R. W. Lindquist. See their article, 'Maximal Analytic Extension of the Kerr Metric', *Journal of Mathematical Physics,* vol. 8, no. 2, February 1967, pp. 265–81. See also an interesting commentary in Chapter 7 of William J. Kaufmann's *Relativity and Cosmology.*

Chapter 3: *Into the Whirlpool*

1 A. Einstein and N. Rosen, 'The Particle Problem in the General Theory of Relativity', *Physical Review,* vol. 48, July 1st, 1935, pp. 73–7. See also, A. Einstein and E. G. Straus, 'The Influence of the Expansion of Space on the Gravitational Fields Surrounding the Individual Stars', *Reviews of Modern Physics,* vol. 17, nos 2 & 3, April–July 1945, pp. 120–4. Rees, Ruffini and Wheeler characterize this latter paper as describing the 'Swiss cheese Universe'; see p. 217 of their *Black Holes, Gravitational Waves and Cosmology.*

2 An interesting article on the mechanics of whirlpools appears in the 1967 *Encyclopaedia Britannica,* vol. 23, p. 477.

3 Edgar Allen Poe, 'A Descent into the Maelstrom', from vol. 2 of his *Works,* 10 vols (Colonial Company, New York and Pittsburgh, 1894).
4 The black hole rotates at slightly more than 400 million m.p.h. (640 million kilometres per hour) if it makes 1,000 complete revolutions per second. Its diameter will be 37 miles (59•3 kilometres); 1,000 revolutions per second is therefore equal to 116,000 miles per second.
5 Schwarzschild's equation predicts that the diameter of a black hole will be $4GM/c^2$, where G is Newton's gravitational constant of $6{\cdot}67 \times 10^{-8}$, and c is the speed of light, at 3×10^{10} centimetres per second (670 million m.p.h.). Taking 1 Solar mass as 2×10^{33} grams (2×10^{27} tons), we compute the diameter of a black hole of 10 Solar masses as follows:

$$\frac{4GM}{c^2} = \frac{4 \times 6{\cdot}67 \times 10^{-8} \times 10 \times 2 \times 10^{33}}{(3 \times 10^{10})^2}$$
$$= 5{\cdot}9 \times 10^6 \text{ centimetres} = 59{\cdot}3 \text{ kilometres}$$
$$= 37 \text{ miles.}$$

The diameter of the vertical strip will be a minimum of one hundredth of the diameter of the disc of the black hole, about 590 metres (640 yards).
6 There appears to be a maximum limit to the mass of an object which can travel unscathed through the disc-edge of the black hole. This may be caused by reasons other than the limited width of the disc-edge. See, M. Simpson and R. Penrose, 'Internal Instability in a Reissner-Nordström Black Hole', *International Journal of Theoretical Physics,* vol. 7, no. 3, 1973, pp. 183–97. 'Internal instability' here means that if a large object of, say, planetary mass, attempts to enter the black hole, it will be crushed. The inner event horizon will, so to speak, 'flip over' and behave like a Schwarzschild singularity. But a relatively tiny object like a spaceship of, say 3,000 tons (3,000 million grams) will not encounter this difficulty in a black hole of ten Solar masses.
7 G. Robinson, 'Hypertravel', *Listener,* December 17th, 1964, pp. 976–7.
8 F. J. Tipler, 'Rotating Cylinders and the Possibility of

Global Causality Violations', *Physical Review D*, vol. 9, no. 8, April 15th, 1974, pp. 2203–6. Tipler developed his work from the solution of some of Einstein's equations by the Scottish mathematician W. J. van Stockum in 1936. Van Stockum's article, in retrospect a most important one, was, 'The Gravitational Field of a Distribution of Particles Rotating About an Axis of Symmetry', *Proceedings of the Royal Society of Edinburgh*, vol. 57, 1936–7, pp. 135–54.

9 B. Carter, 'Global Structure of the Kerr Family of Gravitational Fields', *Physical Review*, vol. 174, no. 5, October 25th, 1968, pp. 1559–71. See also Carter, 'Complete Analytic Extension of the Symmetry Axis of Kerr's Solution of Einstein's Equations', *ibid*, vol. 141, no. 4, January 1966, pp. 1242–7; also, 'Axisymmetric Black Hole Has Only Two Degrees of Freedom', *Physical Review Letters*, vol. 26, no. 6, February 8th, 1971, pp. 331–3. I should add that extensive general bibliographies of various papers on black holes are given in the booklet, *Black Holes (1970–74)*, published by the Institution of Electrical Engineers, and also in pp. 1221–54 of Misner, Thorne and Sheeler, *Gravitation*.

Chapter 4: *The Forbidden Circle*

1 Carter, 'Global Structure of the Kerr Family of Gravitational Fields', *Physical Review*, vol. 174, no. 5, October 25th, 1968, pp. 1559–71.

2 From Edward Bulwer-Lytton's poem 'Orval', which he wrote under the pseudonym of Owen Meredith.

3 The Einstein equation in question states that the slowing of time in a moving vehicle equals the pace of time which it records when stationary, multiplied by:

$$\sqrt{\left(1-\frac{v^2}{c^2}\right)}$$

where v is the speed of the vehicle and c is the speed of light in the vacuum of space. We can see from this equation that when v is of negligible value, the slowing of time inside the vehicle is so slight as to be undetectable. But when v mounts towards c, the slowing of time becomes ever greater. When v equals c, the formula above amounts to zero. In other words, time inside the vehicle stops altogether. If v were greater than c (that is to say, if the vehicle was travelling faster than light) v would have a negative value. Time inside the vehicle would therefore be running backwards.

4 The first of these limericks was written by E. Reginald Buller, of the University of Manitoba. The second is the work of J. H. Fremlin, a physicist at Birmingham University. See Fremlin, 'Newton, Relativity and Free Will', *University of Birmingham Review,* autumn 1966, pp. 42–8.

5 I have culled some of these short stories about time travellers from a collection by Martin Gardner. See his 'Mathematical Games' column, *Scientific American,* May 1974.

6 H. Everett, '"Relative State" Formulation of Quantum Mechanics', *Reviews of Modern Physics,* vol. 29, no. 3, July 1957, pp. 454–62. Everett remarked with disarming ingenuousness in a footnote to his paper that nobody in one 'relative state' could ever be aware of anyone in another. No observer could ever be aware of the 'branching' process. He then added:

> Arguments that the world picture presented by this theory is contradicted by experience, because we are unaware of any branching process, are like the criticism of the Copernican theory that the mobility of the Earth as a real physical fact is incompatible with the common sense interpretation of nature because we feel no such motion. In both cases the arguments fail when it is shown that the theory itself predicts that our experience will be in fact what it is.

See some interesting remarks on Everett's paper in Martin Gardner's article on time travel cited in Note 5 above.

7 It is a widespread view among historians that the war might never have been fought if Hitler had died before

Germany became committed to aggression. See, for example, Joachim Fest, 'Thinking about Hitler', *Encounter*, September 1975, pp. 81–7.

8 Quoted by Gardner, op. cit.

9 Gardner, *Relativity for the Million*, pp. 43–6.

10 Einstein, *The Meaning of Relativity*, p. 26. One fascinating paper that discusses the provability of the simultaneity of distant events in space, by means of a thought experiment involving "fundamental observers," (i.e., observers who would miraculously be able to observe all points in the universe at once), is H.S. Murdoch's "Recession Velocities Greater than Light," *Quarterly Journal of the Royal Astronomical Society*, Vol. 18, pp. 242–7, 1977.

Chapter 5: *The Arm of Orion*

1 Thomas B. Aldrich (1836–1907), 'Sonnet: Miracles'.

2 There is a large amount of literature on interstellar dust. In particular I would recommend Hans Rohr's illustrated book, *The Radiant Universe*. The following articles are also most useful: G. H. Herbig, 'Interstellar Smog', *American Scientist*, vol. 62, March–April 1974, pp. 200–7; P. S. Wesson, 'A Synthesis of our Present Knowledge of Interstellar Dust', *Space Science Reviews*, vol. 15, 1974, pp. 469–82; A. N. Witt and C. F. Lillie, 'Diffuse Galactic Light and the Albedo of Interstellar Dust in the 1500 Angstrom to 4250 Angstrom Region', *Astronomy and Astrophysics*, vol. 25, 1973, pp. 397–404.

3 I have taken this excellent analogy from Rohr's *The Radiant Universe*, p. 31.

4 See Notes 9 and 14 below.

5 See, for example, R. Penrose, 'Black Holes', *Scientific American*, May 1972, pp. 38–46.

6 For details of this book, see the Bibliography under Sandage.

7 Photographs of the spiral galaxies, whose clearest specimens are known as Type Sb in Edwin Hubble's classification, are shown on pp. 12–25 of Sandage's *Atlas*. Perhaps the clearest of all these pictures is the galaxy Messier 81 (N.G.C. 3031), occupying the whole of p. 19. It may well be almost identical to our own Galaxy in age, mass and structure. But this could be equally well said of many other

Sb galaxies; M81 is the best example because the photograph has been so beautifully enlarged. Anyone in Britain unable to see the *Atlas* will be able to find a copy in the library of the Royal Astronomical Society, at Burlington House, Piccadilly, London.

8 The case for believing that another Ice Age is imminent within a few centuries is given by John Gribbin in *Our Changing Climate* (Faber, London, 1975). The contrary view, namely that our knowledge of the Earth's atmosphere is much too limited to justify any such prediction about the future climate is given by the Director of the Meteorological Office in Britain, Dr Basil Mason, in a fierce review of Gribbin's book in the *New Scientist,* vol. 68, no. 971, 1975, pp. 175–6. My own researches have confirmed my view that Mason is right, and that the evidence is insufficient to make any prediction about the climate of the next millennium. See A. Berry, 'Of Ice and Men', London *Sunday Telegraph,* June 29th, 1975.

9 W. H. McCrea, Ice Ages and the Galaxy', *Nature,* vol. 255, June 19th, 1975, pp. 607–9. In another paper in the same journal (vol. 257, October 30th, 1975, pp. 776–8), under the heading 'Galactic Dust Lanes and Lunar Soil', two astronomers, J. F. Lindsay and L. J. Srnka, support McCrea's data, inferring a density of the inner compression lanes of any one of the Galaxy's spiral arms as about 10^{-19} grams per cubic centimetre ($3{\cdot}6 \times 10^{-21}$ pounds per cubic inch). See also, J. Gribbin, 'Dusty Embrace of the Milky Way may Explain Ice Ages', *New Scientist,* June 26th, 1975, p. 695.

10 See an excellent summary of this theory of Ice Ages in Arthur Beiser's *The Earth* (Life Nature Library, New York, 1962 and 1964), pp. 162–3.

11 These figures of 10^{20} grams (10^{14} tons) in a spherical volume region of 10^{40} cubic centimetres ($3{\cdot}37 \times 10^{24}$ cubic miles) surrounding the Sun with a radius of 1 Astronomical Unit ($1{\cdot}5 \times 10^{13}$ centimetres = 93 million miles) are reached as follows: The approximate average density of space in the Orion Arm, 10^{-19} grams per cubic centimetre ($3{\cdot}61 \times 10^{-21}$ pounds per cubic inch) is multiplied by $\frac{4}{3}\pi 10^3$, where r equals $1{\cdot}4 \times 10^{13}$ centimetres. This volume therefore equals $\frac{4}{3}\pi \times 3{\cdot}4 \times 10^{39} = 1{\cdot}4 \times$

10^{40} cubic centimeters = $3{\cdot}4 \times 10^{24}$ cubic miles. Multiplying this by the density, we have $1{\cdot}4 \times 10^{40} \times 10^{-19} = 10^{21}$ grams = 10^{15} tons. By comparison, the mass of the Earth is 6×10^{27} grams (6×10^{21} tons), and the mass of the Sun is 2×10^{33} grams (2×10^{27} tons).

12 D. A. Allen and K. M. Merrill, 'Haro 13a: A Luminous, Heavily Obscured Star in Orion?', *Monthly Notices of the Royal Astronomical Society,* vol. 173 (Short Communications), 1975, pp. 47–50.

13 See, for example, P. G. Manning, 'Ferric Oxide (Alpha-Haematite) in Interstellar Dust', *Nature Physical Science,* vol. 245, October 1st, 1973, pp. 72–3. Also Manning, 'Origin of Broad Interstellar Feature at $1{\cdot}6\ \mu m^{-1}$', *Nature,* vol. 255, no. 5503, May 1st, 1975, pp. 40–1. Also, W. W. Duley, 'Fluctuations in Interstellar Grain Temperatures', *Astrophysics and Space Science,* vol. 23, 1973, pp. 43–50. As Manning sums up the matter (private communication to the author, October, 1975), 'Cosmic abundances suggest that carbon, silicon, iron and magnesium are the important elements comprising (interstellar) grains'. Hematite, also written haematite, has the formula Fe_2O_3. It is found in many places on Earth, from the iron mines of Lancashire and Cumberland in Britain, to the iron ranges of the Lake Superior district, especially in the Mesabi Range in Minnesota.

14. The Orion Arm is considerably more dense than it would be if it consisted only of hydrogen. In addition to iron and nickel, with their respective atomic weights of 56 and 59, the Arm also contains silicon, oxygen and carbon, with atomic weights of 28, 16 and 12. (Hydrogen has an atomic weight of 1.) The density that we must now assume in the Orion Arm, is between a lower limit of 10^{-19} grams per cubic centimetre ($3{\cdot}6 \times 10^{-21}$ pounds per cubic inch) and an upper limit of 5×10^{-19} gm/cm^3 (2×10^{-20} pounds per cubic inch). Taking McCrea's estimates (see Note 9 above), we multiply these figures in turn by $\frac{4}{3}\pi r^3$, where r = 1 light-year, or $9{\cdot}5 \times 10^{17}$ centimetres ($5{\cdot}9 \times 10^{12}$ miles), we obtain a lower limit of 4×10^{35} grams (4×10^{29} tons), and an upper limit of 2×10^{36} grams (2×10^{30} tons). These figures represent respectively about

100 and 1,000 Solar Masses, since 1 Solar mass is 2×10^{33} grams (2×10^{27} tons). And it is believed that a minimum of only 3 Solar masses is required for the formation of a black hole. I am greatly indebted to Anthony Lawton for his kind help with these calculations.

Chapter 6: *The Astromagnets*

1 Michael Faraday, in a great experimental study which he began in 1845, discovered that all the elements then known (and their alloys) were either ferromagnetic, paramagnetic or diamagnetic. To learn precisely which element is which, it is necessary to consult a so-called 'chart of atomic susceptibilities'. I found one on p. 456 of Richard M. Bozorth's *Ferromagnetism.*

2 From the Savoy Operas, *Patience,* Act 2.

3 The most lucid description I have read of man's attempts to control plasma by magnetic mirrors in order to achieve controlled thermonuclear fusion will be found in Samuel Glasstone's *Sourcebook on Atomic Energy,* pp. 540–61. I strongly recommend Tom Alexander's article, 'Fusion Power Breakthrough', his contribution to the 1972 *Nature Science Annual* (Time-Life Books, 1972), pp. 139–49. See also, T. K. Fowler and R. F. Post, 'Progress Towards Fusion Power', *Scientific American,* December 1966, pp. 21–31; W. C. Gough and B. J. Eastlund, 'The Prospects of Fusion Power', *ibid,* February 1971, pp. 50–64; L. Artsimovich, 'The Road to Controlled Nuclear Fusion', *Nature,* vol. 239, September 1st, 1972, pp. 18–22; and F. Knebel, interview with Melvin B. Gottlieb, 'Work in Progress', *Intellectual Digest,* September 1972.

4 Francis Bacon, the *Novum Organum,* 1620.

5 See A. R. Martin's paper, cited in Note 7.

6 R. W. Bussard, 'Galactic Matter and Interstellar Flight', *Astronomica Acta,* vol. 6, fasc. 4, 1960, pp. 179–94.

7 In addition to Bussard's paper cited above, see A. R. Martin, 'Some Limitations of the Interstellar Ramjet'. *Spaceflight,* vol. 14, no. 1, February 1973, pp. 21–5; G. L. Matloff and A. J. Fennelly, 'A Superconducting Ion Scoop and its Application to Interstellar Flight'. *Journal of the British Interplanetary Society,* vol. 27, 1974, pp. 663–73;

and, A. Bond, 'An Analysis of the Potential Performance of the Ram Augmented Interstellar Rocket', *ibid,* same issue, pp. 674–85.

8 A point made most vividly in Sagan and Schlovskii, *Intelligent Life in the Universe,* p. 446.

9 As A. R. Martin explains in his paper (Note 7 above),

> There must come a point where acceleration of the vehicle must be reduced to avoid the breakdown of the vehicle structure by the magnetic forces . . . Acceleration is still possible [beyond a certain cut-off point] . . . but the level must be reduced to compensate for the added strain produced by the magnetic field with the increasing velocity.

10 For literature on O'Neill's plan, see Note 1 of the Introduction.

11 Eros, and asteroids of similar size, are obviously very jagged objects. Beyond their generally agreed roughly cylindrical shapes, there is a good deal of uncertainty. The Arizona astronomer Ben Zellner ('New Findings about Eros', *Sky and Telescope,* vol. 50, no. 6, December 1975, pp. 376–9) calls it a 'stubby cylinder with rounded ends, with the longest dimension 2·3 to 3 times the shortest'. Van Nostrand's *Scientific Encyclopedia* (3rd ed., 1958) says on p. 134 that Eros is 'brick-shaped' and that this is 'definitely proved'. On p. 612 of the same edition of the encyclopedia, Eros is described as 'dumb-bell shaped'. Gunter T. Roth, in his *The System of Minor Planets,* pp. 77–8, likens Eros to a cigar ('a drawn-out ellipsoid') and a boiler (again, a 'cylinder with rounded ends'). So one must take one's choice between a brick, a dumb-bell, a cigar and a boiler. The only thing suggested by all these terms is an oblong shape. See also, E. F. Helin and E. M. Shoemaker, 'Discovery of a Minor Planet', *Astronomy,* vol. 4, no. 6, June 1976.

12 The classic work on exploiting the asteroids is still Dandridge M. Cole and Donald W. Cox, *Islands in Space,* Chapters 9–13 and Chapter 15.

13 K. S. Thorne, 'The Search for Black Holes', *Scientific American,* December 1974, pp. 32–43.

Chapter 7: *The Politicans*

1 T. H. Huxley's evidence to the (British) Select Committee on Scientific Education, 1868. Quoted in Sir Eric Ashby's *Technology and the Academics*, pp. 35–6.
2 C. P. Snow, *The Two Cultures and the Scientific Revolution.*
3 Dr Black is in the right in this dispute, although his definitions are extremely rigid. To an electrical engineer, the phrases 'pilot controller' and 'master controller' are indeed synonymous. See, for example, *Chambers's Dictionary of Science and Technology* (1971), pp. 731 and 892. The *Random House Dictionary of the English Language* (Unabridged ed., 1966) gives a full account of the almost contradictory applications of the word 'pilot'.

Chapter 8: *The Other End of the Labyrinth*

1 The South Korean national flag is pictured in colour in the article on flags in vol. 9 of the 1967 *Encyclopaedia Britannica.* See an interesting passage on the Yin-Yang symbol in Martin Gardner's *The Ambidextrous Universe*, pp. 234–7. The same symbol appears, in line-form only, on every American baseball.
2 The Golden Rectangle and many symmetrical and other mathematical shapes that have a natural beauty are discussed in H. E. Huntley's fascinating book, *The Divine Proportion* (Dover, New York, 1970).
3 Poem, 'The Perils of Modern Living', by H. P. Furth. *New Yorker*, November 10th, 1956, p. 56. Reprinted by permission; © The New Yorker Magazine, Inc.
4 This anonymous limerick was first quoted in Charles Howard Hinton's *Scientific Romances* (London, 1888). See also Gardner, *The Ambidextrous Universe*, p. 250.
5 P. C. W. Davies, *The Physics of Time Asymmetry.*
6 The classic paper that pointed the way to the solution of the Ozma Problem, or, in scientific terms, brought about the 'fall of parity' and which won its authors the Nobel prize in physics, was Chen Ning Yang and Tsung Dao Lee, 'The Question of Parity Conservation in Weak Interaction', *Physical Review*, October 1st, 1956, pp. 254–8. Their work was followed in the same winter by a decisive experiment with cobalt-60 by Madame Chien-Shiung Wu,

a professor of physics at Columbia University, New York. The late John Campbell wrote afterwards in an editorial in *Analog Science Fiction* that it might have been Eastern cultural interest in symmetry which predisposed three Chinese-born scientists to seek out a natural example of asymmetry.

7 This estimate is one of the consequences of Joseph Weber's apparently successful attempts to detect gravitational radiation. His results indicated that the Galaxy may be losing mass at the approximate rate of one Solar mass per day. See J. Weber, 'Gravitational Radiation Experiments', *Physical Review Letters,* vol. 24, February 9th, 1970, pp. 498–50. Also Weber, 'Anisotropy and Polarisation in the Gravitational-Radiation Experiments', *ibid,* vol. 25, July 20th, 1970, pp. 180–4. See also Weber's letter to *Nature,* vol. 240, Nov. 3rd, 1972, pp. 28–30, 'Computer Analyses of Gravitational Radiation Detector Coincidences'. For an excellent layman's explanation, see Tom Alexander, 'Mystery of the Gravity Waves', *Nature/Science Annual* (Time-Life Books, 1971), p. 117. Also G. B. Field, M. J. Rees, and D. W. Sciama, 'The Astronomical Significance of Mass Loss by Gravitational Radiation', *Comments on Astrophysics and Space Science,* vol. 1, 1969, p. 187. And, D. W. Sciama, 'Is the Galaxy Losing Mass on a Time Scale of a Billion Years?', *Nature,* vol. 224, December 27th, 1969, pp. 1263–7. Hjellming's letter (see Note 9 below) and the papers cited in Note 10 are also interesting on this point.

8 Sir James Jeans, *Astronomy and Cosmology* (London, 1928).

9 R. M. Hjellming, 'Black and White Holes', *Nature Physical Science,* vol. 231, May 3rd, 1971, p. 20.

10 Hoyle, *Astronomy and Cosmology: A Modern Course,* p. 696.

11 J. V. Narlikar and K. M. V. Apparao, 'White Holes and High Energy Astrophysics', *Astrophysics and Space Science,* vol. 35, 1975, pp. 321–36. This paper was a development of earlier work on white holes by Narlikar and Apparao. See their earlier article 'High Energy Radiation from White Holes', *Nature,* vol. 251, October 18th, 1974, pp. 590–1. Also, J. Gribbin, 'White Holes—A Coming

Fashion?', *New Scientist*, October 23rd, 1975, p. 199; J. Gribbin, 'Retarded Cores, Black Holes and Galaxy Formation', *Nature*, vol. 252, December 6th, 1974, pp. 445–7; Y. Ne'eman and G. Tauber, 'The Lagging Core Model for Quasi-Stellar Sources', *The Astrophysical Journal*, vol. 150, December 1967, pp. 755–66. These last two papers discuss a possible additional source of energy for white holes, that they may be a delayed mini-explosion or 'lagging core' from the original Big Bang which gave birth to the present cycle of the Universe. If this theory is true, it gives more support to the concept of the Einstein-Rosen bridge. But it also brings us into much deeper questions which are outside the domain of this book. See also J. Gribbin, 'Black Holes, White Holes and Wormholer', *Astronomy*, vol. 4, no. 11, Nov. 1976, pp. 23–6.

12 The legend of Theseus comes to us from a brief biography in Plutarch's *Lives*, although Plutarch mentions an earlier version by Philochorus, which appears to have been lost. For modern versions, see Mary Renault's novels, *The King Must Die* (1958), and *The Bull from the Sea* (1962).

Epilogue

1 From a B.B.C. radio interview with John Taylor, Professor of Mathematics at King's College, London, in the autumn of 1975.

2 Simpson and Penrose, 'Internal Instability in a Reissner-Nordström Black Hole'. See Note 8 of Chapter 3.

3 From Ginzburg's 1975 George Darwin lecture, delivered to the Royal Astronomical Society on April 11th, 1975. Reprinted in *The Observatory*, vol. 95, no. 1008, October 1975, pp. 153–61.

4 Robert Bussard's paper is cited in Note 6 of Chapter 6. The italics used here are my own.

Bibliography

(In the case of two or more authors or editors, the book is listed under whichever is first in alphabetical order.)

ALLEN, C. W., *Astrophysical Quantities* (Athlone Press, London, 1973).

ASHBY, Eric, *Technology and the Academics* (Macmillan, London, 1958).

ASIMOV, Isaac, *The Collapsing Universe* (Walker & Co., New York, 1977).

BECKER, W., and G. Contopoulos, *The Spiral Structure of our Galaxy*, International Astronomical Union, no. 38; seminar held in Basle, Switzerland, August 29th–September 4th, 1969 (Reidel, Boston and Dordrecht, Holland, 1970).

BERGAMINI, David, *The Universe* (Life Science Library, New York, 1962, 1967).

BERGMANN, P. G., *The Riddle of Gravitation* (Murray, London, 1969).

BERRY, Adrian, *The Next Ten Thousand Years: A Vision of Man's Future in the Universe* (Cape, London, 1974; Coronet Books, London, 1976; Dutton, New York, 1974; New American Library, New York, 1975).

BOREL, Emile, *Space and Time* (Dover Publications, New York, 1970).

BORN, Mac (ed.), *The Born-Einstein Letters* (Macmillan, London and New York, 1971).

BOVA, Ben, *The New Astronomies* (Dent, London, 1973).

——— *The Analog Science Fact Reader* (Condé-Nast, New York; Millington, London; both 1974).

BOZORTH, Richard M., *Ferromagnetism* (Van Nostrand, Toronto, New York and London, 1951).

BRECHER, Kenneth, and Giancarlo Setti, *High Energy Astrophysics and its Relation to Elementary Particle Physics* (Massachusetts Institute of Technology Press, 1974).

CALDER, Nigel, *Violent Universe: An Eye-Witness Account of the New Astronomy* (B.B.C. Publications, London, 1969 and 1973).

CAMERON, A. G. W., and George B. Field, *The Dusty Universe* (Published for the Smithsonian Astrophysical Observatory by Neale Watson Academic Publications, New York, (1975).

CARMELI, Moshe, Stuart I. Fickler and Louis Witten (eds), *Relativity: Proceedings of the Relativity Conference in the Midwest, held at Cincinnati, Ohio, June 2–6, 1969* (Plenum Press, New York and London, 1970).

CHEMICAL RUBBER COMPANY, Samuel M. Selby (ed.), *Handbook of Tables for Mathematics* (Cleveland, Ohio, 1970).

CLAIBORNE, Robert, and Samuel L. Goudsmit, *Time* (Life Science Library, New York, 1966).

CLARKE, Arthur C., *The Promise of Space* (Hodder & Stoughton, London, 1968; Pelican Books, Harmondsworth, 1970).

COLE, Dandridge M., and Donald W. Cox, *Islands in Space: The Challenge of the Planetoids* (Chilton Books, New York, 1965).

DAVIES, P. C. W., *The Physics of Time Asymmetry* (Intertext Publishing in association with Surrey University Press, England, 1974).

DEWITT, Cecile M., and John A. Wheeler (eds), *Battelle Rencontres: 1967 Lectures in Mathematics and Physics* (W. A. Benjamin, New York and Amsterdam, 1968).

DOLE, Stephen, *Habitable Planets for Man* (Elsevier, New York, 1970).

DURELL, Clement V., *Readable Relativity* (Harper & Row, New York, 1966).

DUVEEN, Anneta, and Lloyd Motz, *Essentials of Astronomy* (Columbia University Press, New York and London, 1966).

EDDINGTON, Arthur, *Stars and Atoms* (Oxford University Press, 1926).

EINSTEIN, Albert, *The Meaning of Relativity*, four lectures delivered at Princeton University, May, 1921 (Methuen, London, 1922; Chapman & Hall Paperbacks, London, 1973).

——— *The Principle of Relativity* (Dover Publications, New York and London, 1952; first published 1923).

ELLIS, G. F. R., and S. W. Hawking, *The Large Scale Structure of Space-Time* (Cambridge University Press, 1973).

GARDNER, Martin, *Relativity for the Million* (Macmillan, New York, 1962; Pocket Books, New York, 1965).

——— *The Ambidextrous Universe* (Pocket Books, New York, 1962; Allen Lane, London, 1967; Pelican Books, London, 1970).

GLASSTONE, Samuel, *Sourcebook on Atomic Energy* (Van Nostrand, Princeton, New Jersey, 1967).

GOLDEN, Frederic, *Quasars, Pulsars, and Black Holes: A Scientific Detective Story* (Scribner, New York, 1976).

GOLDSMITH, Donald, and Donald Levy, *From the Black Hole to the Infinite Universe* (Holden-Day, San Francisco and London, 1974).

GORDON, Mark A., and Lewis E. Snyder (eds), *Molecules in the Galactic Environment* (John Wiley, New York and London, 1973).

GREENBURG, J. M., and H. C. van de Hulst, *Interstellar Dust and Related Topics* (Published for the International Astronomical Union by Reidel, Dordrecht and Boston, 1973).

GRIBBIN, John, *White Holes: Cosmic Gushers in Space* (Belacorte Press, New York, 1977).

HARRISON, B. Kent, Kip S. Thorne, Masami Wakano and John Archibald Wheeler, *Gravitational Theory and Gravitational Collapse* (University of Chicago Press, 1965).

HOYLE, Sir Fred, *The Nature of the Universe* (Blackwell, Oxford, 1960).

——— *Astronomy and Cosmology: A Modern Course* (W. H. Freeman, San Francisco, 1975).

ICHIMARU, S., *Basic Principles of Plasma Physics* (W. A. Benjamin Advanced Book Program, Reading, Massachusetts, 1973).

INSTITUTION OF ELECTRICAL ENGINEERS, *Black Holes 1970–74, Bibliography* (London, 1975).

JASTROW, Robert, *Stars, Planets and Life* (Heinemann, London, 1968).

——— and Malcolm H. Thompson, *Astronomy: Fundamentals and Frontiers* (John Wiley, New York and London, 1972).

JOHN, Laurie (ed.), *Cosmology Now* (B.B.C. Publications, London, 1973).

KAHN, F. D., and H. P. Palmer, *Quasars: Their Importance in Astronomy and Physics* (Manchester University Press, 1967 and 1968).

KAUFMANN, William J., *Relativity and Cosmology* (Harper & Row, New York, San Francisco and London, 1977).

KLAUDER, John R. (ed.), *Magic Without Magic: John Archibald Wheeler* (W. H. Freeman, San Francisco and Reading, 1972).

LAPP, Ralph E., *Matter* (Life Science Library, New York, 1965).

LAWTON, Anthony, and Jack Stonely, *Ceti* (Warner Brothers, New York, 1976).

——— *Is Anyone Out There?* (Warner Brothers, New York, 1974).

LEQUEUX, J., *Structure and Evolution of Galaxies* (Gordon & Breach, New York, London and Paris, 1969).

LEVITT, I. M., *Beyond the Known Universe: From Dwarf Stars to Quasars* (Viking Press, New York, 1974).

LUNAN, Duncan, *Man and the Stars: Contact and Communication with other Intelligence* (Souvenir Press, London, 1974).

LYTTLETON, R. A., *Mysteries of the Solar System* (Clarendon Press, Oxford, 1968).

MASSEY, Harrie, *Space Physics* (Cambridge University Press, 1964).

MEADOWS, A. J., *Stellar Evolution* (Pergamon Press, London and New York, 1967).

MISNER, Charles W., Kip S. Thorne and John A. Wheeler, *Gravitation* (W. H. Freeman, San Francisco and Reading, 1973).

MOORE, Patrick, *Atlas of the Universe* (Mitchell Beazley, London, 1970).

——— and Iain Nicolson, *Black Holes in Space* (Orbach & Chambers, London, 1974).

MOSKOWSKI, Alexander, *Conversations with Einstein* (Sidgwick & Jackson, London, 1973; first published, 1921).

OSTERBROCK, Donald E., *Astrophysics of Gaseous Nebulae* (W. H. Freeman, San Francisco and Reading, 1974).

PAGE, Lou Williams, and Thornton Page, *Stars and Clouds of the Milky Way: The Structure and Motion of our Galaxy* (Macmillan, New York; Collier-Macmillan, London, both 1968).

PARNOV, E. I., *At the Crossroads of Infinities* (Mir, Moscow, 1971).

PEARCE WILLIAMS, L. (ed.), *Relativity Theory: Its Origins and Impact on Modern Thought* (Wiley, New York and London, 1968).

POLGREEN, G. R., *New Applications of Modern Magnets* (Macdonald, London, 1966).

REES, Martin, Remo Ruffini and John A. Wheeler, *Black Holes, Gravitational Waves and Cosmology: An Introduction to Current Research* (Gordon & Breach, New York, London and Paris, 1974).

ROHR, Hans, *The Radiant Universe* (Warne, London, 1972).

ROTH, Gunter D., *The System of Minor Planets* (Faber, London, 1962).

RUSSELL, Bertrand, *The A.B.C. of Relativity* (Allen & Unwin, London; New American Library, New York, both 1958; first published 1925).

SAGAN, Carl, *The Cosmic Connection: An Extraterrestrial Perspective* (Doubleday, New York, 1973; Hodder & Stoughton, London, 1974).

——— and Iosif Schlovskii, *Intelligent Life in the Universe* (Holden-Day, San Francisco, 1966).

SANDAGE, Allan, *The Hubble Atlas of Galaxies* (Carnegie Institute of Washington, 1961).

SHIPMAN, Harry L., *Black Holes, Quasars and the Universe* (Houghton-Mifflin, Boston, 1976).

SHOENBERG, David, *Magnetism* (Sigma Books, London, 1949).

SKLAR, Lawrence, *Space, Time and Spacetime* (University of California Press, 1974).

SNOW, C. P., *The Two Cultures and the Scientific Revolution* (Cambridge University Press, 1959).

SWIHART, Thomas L., *Astrophysics and Stellar Astronomy* (Wiley, New York and London, 1968).
TAYLOR, John, *Black Holes: The End of the Universe?* (Souvenir Press, London, 1973; Fontana, London, 1974).
TERLETSKII, Yakov P., *Paradoxes in the Theory of Relativity* (Plenum Press, New York, 1968).
TOFFLER, Alvin, *Future Shock* (Bodley Head, London, 1970; Pan Books, London, 1971).
UNIVERSITY OF BRUSSELS, *Astrophysics and Gravitation: Proceedings of the 16th Solvay Conference at the University of Brussels, September, 1973* (Éditions de l'Université de Bruxelles, 1974).
WHEELER, John A. (ed.), *Geometrodynamics* (Academic Press, New York and London, 1962).
WHITNEY, Charles A., *The Discovery of our Galaxy* (Thames & Hudson, London, 1972).
WHITROW, G. J., *What is Time?* (Thames & Hudson, London, 1972).

Index